Miracle in the Void

Meredith
Miller/95

MIRACLE IN THE VOID

Free Energy, UFOs and Other Scientific Revelations

by

Brian O'Leary

Illustrations by Meredith Miller
Photographs by the Author

Kamapua'a Press

Miracle in the Void:
Free Energy, UFOs and Other Scientific Revelations

Printed in the United States of America on acid-free recycled paper.

Second printing March, 1996

Kamapua'a Press
1993 S. Kihei Road
Suite 21-200
Kihei, Hawaii 96753

Cover design by Yevgeniy Epshteyn, sourced from *The Resurrection of Gaia* by Meredith Miller.

O'Leary, Brian, 1940-
Miracle in the void : free energy, UFOs and other scientific revelations / by Brian O'Leary ; illustrations by Meredith Miller, -- Kihei, Hawaii : Kamapua'a Press, 1996.
p. cm.
Includes bibliographical references and index.
Preassigned LCCN: 95-78998
ISBN 0-9647826-0-X

1. Force and energy. 2. Science-Philosophy. I. Title

QC73.064 1996 530
QB195-20376

To Gaia, our Earth
for her nurturing patience
and beauty

And to the courageous
inventors and researchers
who have tapped into the void
to bring forward a healing Light

Acknowledgments

Nobody has helped serve me better than my spiritual partner–artist Meredith Miller. She has endured my three–year process of grief that led to creating this book. She has been inspirational in ideas and tolerant of my mood swings. I am forever grateful to her.

I also thank Robert Perry, Pat Proud, Michael Brein, Marlene Oaks, Marsha Casavant, Lorraine Katena, Donna Rae, Jeane Manning, Jonathan Eisen, Paul Von Ward, Dan Drasin, Robert Gould, Sparky Sweet, Toby Grotz, Pat Bailey, Bill Beierwaltes, Maury Albertson, Bob Siblerud, Patrick and Diane McNeal, Paramahamsa Tewari, Shiuji Inomata, John Hutchison, Yin Gazda, Hal Puthoff, Don Kelly, Kirk Thatcher, Whitley Strieber, Bruce DePalma, Andrew Mount and so many countless others for helping me through this most challenging project.

Finally, we thank Yevgeniy Epshteyn for the cover design, sourced from *The Resurrection of Gaia* by Meredith Miller.

Contents

Part I
Death in the Old Paradigm

Part II
Resurrection in the New Paradigm

Prologue

Gaia

Gaia a forgotten prayer
Tossed on the wind
The ancient ones know your glory
Spoke of it with ritual and reverence

Spun from star dust
A brew of fire and ice
Wrought of lava and liquid
Roughly hewn
Delicate in balance
A lapis lazuli orb
Set in a web of nimbus clouds
The blue jewel of the heavens

In your smoldering core
Do you plot retribution
While we shroud you in poisonous vapors
And worship our gods of garbage and greed
Pillaging with deluded dreams of dominion
Carving treason through your forests
Exiling your beasts of land and sea
To early extinction and modern fable
These are crimes of our dispassion
Your heart trembles earthquakes
At our sacrilege

In your molten belly
Are you forging new mountains
That will rumble
Bursting through your ruined surface
To topple your tenuous cities
Will your oceans rise up
To wash you clean of civilization
Will you sculpt new continents
With our bones and ashes
Will birds hail the dawn
With lilting songs of our demise

Suspended in Cimmerian skies
Sister Moon smiles cool
The wolf howls his greeting
Across dark infinity
And we wash up against
The shores of oblivion
Our bones like sea shells
Tossed on the sand
The age of man
A forgotten dust on the wind...

–Catherine Hayes

Forewords

Miracle in the Void is a unique contribution to the understanding of a new scientific paradigm. Dr. Brian O'Leary combines consciousness research with zero–point energy (*empty* space is not empty) and novel concepts of electrical power generation. Are you ready for good news? While other physicists shy away from the implications of new physics, he bluntly tells us we can replace the oil–based world economy and renovate our nation–states as well as clean up our environment. This ambitious work is a good read; he weaves in diary entries that reveal the emotions a science insider faces after expanding his or her *box* of knowledge. O'Leary is compassionate toward establishment scientists who avoid taking that step.

If you've already said farewell to the old paradigm, faced the fact that fossil–fuel pollution was unnecessary, and worked through the stages of anger and grieving which O'Leary so clearly describes in Part I, jump to Part II of *Miracle in the Void* for an enlightening treat. Dr. O'Leary's science background and globe–spanning travels help to synthesize theoretical and experimental work in the emerging fields of "free energy" and consciousness research. The photographs of inventors he interviewed are works of art in themselves. His book is for everyone who wants beneficial changes on our planet.

–Jeane Manning, author of *The Coming Energy Revolution,*(Avery Books, 1995)

Do you remember vaguely hearing somewhere that we're depleting our fossil fuels, overpopulating and destroying our planet? Has all this stopped, simply gone away? In our quest to survive, a monumental revolution in consciousness —no less than a revolution in the way we think—is coming. The coming colossal and catastrophic changes to the whole of society and all its institutions will surely affect us all. Life as we know it will never be the same.

Brian O'Leary in his brilliant breakthrough book *Miracle in the Void* pulls no punches but instead drags us kicking and screaming to awareness of the coming inevitable changes, forcing us to face our denial, our anger, our despair and, finally, showing us with great compassion by his own life's example how we may—indeed, how we *must*—proceed through the steps of grieving our losses of the old ways of thinking and surviving in order to prepare to accept the new. But the news is not all bad! Energy and abundance for all not only is possible—it is ours just for the asking!

Dr. Michael Brein, Consultant in Social Psychology and
Ambassador-at-Large for the Mutual UFO Network

This we know--the earth does not belong to man, man belongs to the earth. All things are connected like the blood which unites one family. Whatever befalls the earth befalls the sons of the earth. Man did not weave the web of life; he is merely a strand in it. Whatever he does to the web, he does to himself.

–attributed to Chief Seattle

Preface

This book completes my new science trilogy, an exploration which has taken me all over the world many times for a decade. My quest has been to witness demonstrations and experiments on concepts that transcend our current scientific worldview (paradigm) of reality. The evidence presented to me has been so dramatic, it confirms a shift in the near future to a global culture that will not resemble the one we have now.

My earlier books *Exploring Inner and Outer Space* and *The Second Coming of Science* look at the UFO/abduction phenomenon, the Mars face, crop circles, the near-death experience, reincarnation, psychokinesis, mind-over-matter, healing, Earth energies, and the latest theories on physics and consciousness. *Miracle in the Void* focuses on our potential to tap the abundant "zero-point" (free) energy of space.

Much of this trilogy is connected with two organizations I have cofounded: the International Association for New Science (IANS) and the Institute for New Energy (INE). Our purpose is to gather scientists and informed lay-people who are not afraid to use the scientific method in helping make a paradigm shift outside the restricted box of Western science.

My initial motivation for this project was to report the latest on free energy developments, based on eyewitness accounts (including mine) of proof-of-concept. But when I really began to reflect on the implications of all this, I saw that I had a tiger-by-the-tail, because all existing energy sources would become obsolete. The economy would have to change. There would be feelings of grieving the old culture.

The reality of being able to extract cheap, clean and abundant energy from the vacuum has only recently been acknowledged by mainstream physicists in the peer-reviewed literature, and it is being developed now in Japan and elsewhere. My technological forecast on this one is radical. Watch out for what's coming!

My perception that we could replace the polluted, depleted fossil fuel economy soon with a free energy era is not a new one. It has been with us for more than a century since the time of Nikola Tesla. It seems incredible that nobody in power or with substantial money has embraced this possibility or has taken sufficient time to look at it. Having been an astronaut in the Apollo program, I have seen goal-setting and vision at their best. We can do the same for free energy. This book deals with both the suppression and the promise of free energy and other scientific revelations.

There is another dimension to all this, the needs of Gaia, Goddess of Planet Earth. My spiritual partner Meredith Miller spent the year 1994 painting *The Last Supper of Gaia* (see photo), which depicts the phases of grief and transformation we will need to go through to embrace the new paradigm. Each disciple at the table represents a particular phase: denial, anger, bargaining, depression, acceptance, creativity, enlightenment, empowerment, joy and transcendence. These also comprise the flow of chapters in this book.

Most of us are still stuck in denial about possibilities, and recent events such as the tragic Oklahoma City bombing suggest that we as a society are going through a transition from the denial to the anger phase.

I believe we are entering a Consciousness Revolution, based on an emerging sacred science, which has far greater implications than anything we can imagine. I hope this book will make that clear. We all seem to be very powerful co–creators of this universe, but the resistance to change is also staggering.

For those of you who may not have read my other books or who may need more information or context of where we will be going here, I refer you to the Appendices for further reading before you enter the stream of my story. Both Appendices also provide some detailed background information about free energy and its implications.

If you run into terms which elude you, I suggest you look at the Glossary, also toward the back. References and Resources sections will help you get into the literature in this rapidly growing field.

Miracle in the Void is divided into two parts. The first explores the phases of grief and suppression, with myself as a case study. The second proposes solutions. If you feel you have gotten your negativity out of the way and are ready for the new paradigm, I recommend you skim the first part and go immediately to the second part, which reveals truly good news.

By the way, it is snowing here now on the chaotic eve of our move to Maui.

Brian O'Leary
Ashland, Oregon
June 6, 1995

UPDATE

"If Thomas Edison invented electric light today, Dan Rather would report it on CBS News as 'The candlemaking industry was threatened.' You would get a downer. You would get a story explaining that electricity kills. There would be interviews with five candlemakers. At least three politicians would pass a bill banning electricity." -Newt Gingrich

While our House Speaker may not be aware of the new Edisons described in this book, we are seeing several encouraging developments on the free energy front.

American inventor James Patterson has patented and demonstrated a fuel cell, based on cold fusion technology, which repeatedly produces over a hundred times more energy than appears to be consumed. On February 7, 1996, Nightline and Good Morning America gave this development credible coverage. Perhaps the mainstream media blackout on free energy is coming to an end.

I have further developed the concepts prescribed in Chapters 9 and 11 in a paper "The Relationship Between Consciousness and the Zero Point Energy" presented in April, 1996, at the Third International Symposium on New Energy held in Denver. Here I explore quantitative equivalence principles for relating experiments on psychokinesis with those on free energy.

We have also established within IANS an Academy for New Energy which will publish a peer-reviewed journal, provide test equipment for researchers and hold conferences. A new research and education facility, the Global Power Station, Maui, is also being birthed. You will be apprised of these and further updates if you are on our mailing list or if you wish to consult our new worldwide web home page (http://www.maui.net/~kamapuaa/).

Brian O'Leary

Kihei, Maui, March 11, 1996

PART I

DEATH IN THE OLD PARADIGM

1

The Relationship Between the Ecological Crisis, Extraterrestrial Intelligence and Free Energy

"I also get a great deal about the destruction of the Earth's ecology which is conveyed by the beings (aliens). And I get that in case after case."

–John Mack, Professor of Psychiatry, Harvard University, and author of Abduction

Puttaparti, India, February 15, 1994

"You look better," Sai Baba said to me, "*much* better!"

"Yes, I feel better," I grinned.

He smiled back. Grasping my hands into his, he looked into my eyes and gazed down my body. "You lost weight. Much, much healthier, yes?"

He pushed me over toward the men's side of the small crowded room with about twenty of us selected for a private interview in his ashram palace. "Sit down, sit down," he said.

And so started my third audience with the great swami, worker of miracles, considered by his fellow Indians as a living national monument.

For three years in a row I was co–leader of a group of a dozen travelers on a "Saints and Miracles" tour of India. Sai

Baba had taken a liking to me, perhaps because of my credentials as an ex–astronaut. Each year he had picked me out of audiences, and invited my groups into his private quarters for one–to–two hour interviews. Several times during each visit the swami had materialized things a few feet away from our unbelieving eyes. He also had begun some extraordinary healings experienced by a few of our group members.

I am not a devotee of Sai Baba or of any other guru, but I felt kindred to this man of great gifts. During my previous visits one and two years ago he had already manifested a beautiful ring and a watch for me while sharing his lighthearted, childlike wisdom.

At the close of this morning's private audience he handed out to us his packets of *vibuti*, the holy ash, and then walked briskly through our gathering back out toward the palace grounds where ten thousand devotees were awaiting another glimpse of the swami.

He then turned around at the door to the room and said directly to me, "I will see you this evening."

Feeling divided loyalties between the members of the group and of fulfilling Sai Baba's invitation, I walked out of the palace barefoot into the boiling midday heat among the throngs. My main concern at this point was, how can I conduct a debriefing session with the group about our encounter, eat, shop and nap? I had only five hours before I would once again return to be separated from the crowd by Sai Baba's familiar hand gesture and joyful declaration, "Moon man, moon man, you come!"

During the day I wondered what he meant by "this evening". Early, just after the afternoon darshan or much later? Did the group want to depart earlier, possibly leaving me stranded at the ashram, accessible only by private car or bus along a bumpy three–hour drive to Bangalore to catch two successive red–eye flights back to our remote home? Did he mean to invite the group as well? What should I ask him if I were to be granted a truly private, personal session with him?

On most days, Sathya Sai Baba holds two *darshans*, or blessings, with crowds of silent devoted thousands. He comes strolling out for a few minutes, acknowledging individuals, gathering their letters containing prayers, and dazzling the onlookers by producing *vibuti* from his hand. A front row seat is coveted, reserved only for VIPs—and for others by lottery. Security was tight this year, with metal detectors at the gates because of an incident that previous June of an attempted assault on him. He barely managed to escape, while his guards killed the assailants.

In eager anticipation of my own one–on–one session with the swami, I set out once again with our group for the afternoon *darshan* in the searing sun. This time the guru picked just me from our group, so I joined him with another group from Hong Kong for about an hour. Then he looked at me and told me to get up and walk with him into a dark back room of the palace. He sat on a small throne and asked me to sit on a pillow to his right, below and physically close to him. He gazed into my eyes lovingly and intensely for a few moments.

"What would you like to ask me?" he began.

"I'm—just—glad to be here. My main question regards a realization I'm having. Just before I came here to see you, I visited with Dr. Paramahamsa Tewari in Karwar."

"Yes, I know him," he said.

"Yes, he admires you very much and has a picture of you on the wall of his laboratory. As you may know he is chief project engineer for the Kaiga nuclear power plant construction project, the largest one in India. He has also developed a free energy machine. Do you know about free energy?"

"Yes, yes," he smiled knowingly.

"Well, my insight is this," I continued. "I saw Tewari create energy out of nothing. Then I have just seen you create matter out of nothing. That must mean that nothing is something. It's our consciousness, right?"

"Yes, yes! You got it. That's it!"

In a flash of illumination reserved for special moments like this, I suddenly understood the reason for my quest which has taken me to the far corners of the Earth. In a series of dramatic demonstrations, I observed the sacred cows of materialism and our outmoded ways of producing energy being sacrificed by two exceptional, gentle men from India.

The coming paradigm shift became a clear vision to me. We scientists are going to have to consider once again the ether, that invisible substratum of reality that underlies the creation and destruction of matter and energy. We will need to recognize that this may be the basis for the inner and outer revolution we are beginning to birth.

We seem to be dealing with two simultaneous realities, the one we are told about by conventional scientists and another that is transcendent and multi-dimensional—a broader and more sacred science which really represents what we really know up to this point as a culture. I have found these realities to be in deep conflict. I have also discovered that only the new science would have the power and potential to save a dying Gaia, our beautiful planet Earth.

My Perceptions as a Scientist and Ecologist

I am a scientist, a seeker of truth.

I view myself as an explorer–observer–participant in a quest to better understand the essence of our reality. I also see myself as an early warning system for what our planet may expect in the future. I occasionally feel like a rev–lutionary who is angry about our shortsighted, suppressive and ecologically destructive culture. And I am eager to help create those social structures that will facilitate a new world view; one that will support a sustainable global future.

In the first few chapters of this book, I make little effort to hide my feelings. This is not what one would expect of a scientist. My reason for doing so is to use myself as a case study to examine the grieving process we are all experiencing around the coming shifts. Emotions are an important aspect of what is happening.

Choosing a career in what we call science seemed to be a wise choice for me as a young person, particularly given my childhood interest in space exploration, which I predicted would become a reality in my lifetime. I ascended rapidly through the educational system to receive a Ph.D. at Berkeley in astronomy in 1967. Then, at the ripe young age of 27, I became the nation's second youngest astronaut, destined to go to Mars, before NASA canceled that program.

During the late 1960s and early 1970s, I had enjoyed, as an insider, a brief decade of the extraordinary growth of the American space program. I had built a solid, safe career with the awards of prestige and financial security.

But, as the 1970s began to close, while holding a faculty position in the physics department at Princeton University, I began to have some experiences that appeared to violate the "laws of nature" that I had so revered and had taught as my gospel. A remote viewing experience, a near-death experience, a mind-over-matter healing of an "in-urable" knee, all led me into a new territory which none of my scientific colleagues seemed to want to enter.

Gradually, as I came into more knowledge of the potential primacy of consciousness over the material world, my old world view began to fall apart. I realized I had to abandon my outmoded ideas that the physical universe was merely the result of a mindless cosmic Big Bang, and that we are inert and mortal victims of the deterministic exchanges of matter and energy based on an event that happened 15 billion years ago. In this old view, we saw God as some impersonal, detached agent who may not even exist any more. But now I began to see the universe in a different way—as being interconnected and full of creative forces that are continually breaking the so–called immutable laws that had seemed to hold us in their grip.

The UFO Phenomenon as My Catalyst

As I sifted through mountains of evidence I gradually had to abandon my old view that those who investigated UFO and "perpetual motion" phenomena represented an amateur and sometimes lunatic fringe not worth dealing with and potentially disruptive of my impeccable career as one of the high priests of our culture. Along with all the others, for convenience, I tabled inquiries into these matters until I made a clean break with scientific orthodoxy about ten years ago. Then I joined ranks with other researchers into some of the unaccepted aspects of the unknown.

I soon learned about the scientific credibility of UFO contacts, the convincing physical evidence from photographs, multiple–witnessed events, physical traces from landings and on abductees, animal mutilations, crop circles, evidence of artificial structures on Mars, etc.

I began to reflect on the implications of all this. As I got further and further into the physical evidence I began to develop some unsettling questions: Are we being manipulated for the sake of genetic and psychic experimentation? Are we becoming, at least to some degree, assimilated into, altered or conquered by other species? Must our self–image now shift from that of being the bullies on the block to being a bit like laboratory animals? Very humbling thoughts.

Pondering these questions led me to look at other anomalies such as psychokinesis, telepathy, alternative healing, prophecy, near-death experiences, spirit communication and possession, reincarnation reports, crop circles, the face on Mars...and free energy. What surprised me most was, that the more time I spent looking at the evidence, and the more rigorously, the more real these phenomena became. I was often shocked and dismayed about these things, leading to many of the uncomfortable feelings I am expressing here.

I also learned that many of these developments will almost certainly lead to technologies that will enable us to solve our greatest environmental challenges. Modern history has shown us repeatedly that, more than any other factor, our choices of new technologies have marked our greatest changes and shaped our lifestyles. We have touted these inventions as the catalysts to new destinies, whether they be a telephone, rocket ship, satellite or computer chip.

Of the possible new technologies awaiting us, free energy appears to be the most immediately impactful. My other two books in this trilogy report on new science concepts in general. This book focuses primarily on the personal and global implications of using free energy sources that will surely trigger a massive paradigm shift. When this happens, life will never be the same for all of us.

I am also discovering that one does not need to make any metaphysical assumptions or speculations to come to this conclusion. Free energy alone can go a long way to transform our precarious ecosystems. Many mainstream scientists agree that free energy and other "anomalies" do exist. But they see these events as exceptional, esoteric, obscure or transient phenomena not to be concerned about as a major focus of scientific inquiry. After all, you can't bring most of these happenings into your laboratory, so why bother?

There is no case more typical of the collision of paradigm forces than that of John Mack, Professor of

Psychiatry at Harvard University and author of the recent book *Abduction*, which is a clinical study of patients who report they have been abducted by extraterrestrial and/or interdimensional beings. The internal consistency among the abductees' stories has been well–documented by Mack and others, and provides additional evidence that these events have in fact occurred. The evidence also points toward a need for us to focus on the Earth's fragile ecology, during these dramatic times of transformation.

In spite of all this, Harvard recently appointed a "secret" faculty committee that is attempting to oust him from his tenured position. "The convening of the secret panel," writes Daniel Sheehan, Mack's attorney during the initial proceedings, "grossly violates all accepted standards of law and due process recognized by United States Courts which standards were established pursuant to principles enunciated by the American Association of University Professors. And the secret procedures adopted by the committee violate every known principle of academic freedom."

In its draft report, the first finding of the committee was doubt that the UFO phenomenon was real, which was central to their arguments. They had cited only one reference of second–hand information to bolster their own denial of UFOs and abductions. They made it clear that it was "professionally irresponsible" for any academic to give any credence to the phenomenon. The committee gave Mack just a few weeks to defend himself against the charges!

Former engineering school dean Robert Jahn at Princeton University has also had to endure the heat of a faculty committee which largely denied the reality of his positive observations of psychokinesis. Witch hunts are alive and well in the halls of ivy.

In expressing such attitudes of denial and disdain toward their own peers, many of our traditional scientists are missing the boat. By smugly insisting that "extraordinary claims require extraordinary proof" they have set up a scientific double standard that safely allows them to ignore important and exciting new evidence. By this means, many have opted themselves into secure positions as technicians of a confining view of reality that is almost purely mechanistic, reductionistic and deterministic. In a word, it is dead.

History is full of examples of scientific denial, but this time it is particularly poignant. Because of the ecological crisis and our self-awareness crisis, I believe the evidence for aliens, free energy, and alternative healing will lead the way toward the most profound global changes we have ever experienced.

These conclusions follow from my quest for understanding reality over the last fifteen years. This journey has taken me all over the world to seek out gifted individuals concerned with applying the scientific method to validating and characterizing an unfolding world view that transcends the narrow box of inquiry which underlies most of what is called science today.

Implications of Free Energy

It is my scientific opinion that free energy, a technology which has been formerly suppressed and denied, is almost here and could soon become available to all of us. The immediate practical questions then arise: How can we reverse our long, sordid history of secrecy and disdain that has beleaguered inventors and researchers alike? How can our economic system come to grips with this emerging new technology that would supplant the current, environmentally destructive, $2 trillion per year global energy infrastructure, that clearly dwarfs any prior invention for its overall influence—even the automobile, airplane, television and computer? Will the economic system as we know it need to be scrapped? How can we reallocate our priorities? And how can we prevent ourselves from once again abusing a new technology so we don't simply end up with bigger power saws and weapons of even more awesome destruction?

The prospective development of free energy also forces many difficult questions such as: Will we be able to set aside our heavily vested interests to come together as a global village to convert to a free energy economy for a sustainable future? How can we survive changes that are likely to be greater than any the world has ever faced? How can we re–educate ourselves to be empathetic to those who will be displaced in the process? What effect will free energy development have on other prospective new–paradigm events such as UFO/alien revelations and vice versa? How closely related are these two quantum leaps into the future? What specifically will the new world look like and how can we get from here to there?

I base my scientific and factual reporting on an in–depth study of free energy I have carried out over the past two years throughout the United States, India, Japan, Canada, New Zealand and Europe. During this time I have visited inventors and other researchers who have been demonstrating proof–of–concept of what is called "zero–point energy production." I have also participated in two major inter–national symposia of scientists and engineers as well as inventors. I have no doubt that both theory and experiment clearly show that tapping this form of energy is now in hand. The question is no longer whether or not the anomalies are real. The question is, what do we do about it?

I had no better demonstration of proof–of–concept than my observations of Paramahamsa Tewari's unipolar generator in Karwar, India, whose electrical output, above a threshold revolutions–per–minute (RPM), was clearly greater than its input. It becomes easier to envision mass–produced models as generators that could soon replace our centralized electrical powerplants and internal combustion engines.

Free energy can help alleviate the world's nutrition and sanitation problems by permitting low–cost extraction and purification of water for domestic use and for irrigating dry land. If implemented properly, this source of energy could solve most of our environmental problems. Because the underlying physics of free energy and antigravity concepts are the same, this technology is also likely to lead to propulsion systems that will revolutionize transportation on and beyond the Earth. This symbiotic relationship has been noted by many leading theoretical physicists, and may represent the advanced technology apparently demonstrated repeatedly to eyewitnesses of the irregular and mysterious motions of UFOs as well as the top–secret activities around Area 51 in Nevada.

What Is Free Energy?

I define free energy as being that which can be extracted from the quantum fluctuations of space itself, either by machines or by our own consciousness.

Theory predicts that this energy source is so vast, that one cubic centimeter of free space contains enough of it to evaporate all the Earth's oceans. Free energy is not a form of nuclear power or solar power. It is very different from anything we have read about in conventional textbooks. Free–energy technology will probably include cold fusion devices, specially–designed magnetic motors, certain "Tesla" devices, and solid state systems.

There is much more to this story than the potential for free energy machines to supplant our environmentally abusive energy infrastructure. Whether it be a Sai Baba mani–festation, bending a spoon with the mind, or creating our own realities, I will be developing the hypothesis that we ourselves are "free energy devices". I will be examining the thought that we have the potential, as sentient beings, of transforming and transmuting the material world through the energies of our consciousness.

This hypothesis is not foreign to those in New Age and metaphysical studies, but it takes on new meaning when we see the energies of our consciousness in action, in the laboratory and in the field. Examples include the important studies conducted at Princeton University by Robert Jahn and Brenda Dunne and the materializations of Sai Baba which I have witnessed recently on several occasions and have described in *The Second Coming of Science*.

The Consciousness Revolution

We are living in a period of denial of the obvious. During the Copernican Revolution, most people still thought that the world was flat and at the center of the universe. The resulting paradigm shift took place slowly because denial became a tactic with which the establishment attempted to maintain power. One nostalgic vestige of the pre–Copernican paradigm is the Flat Earth Society of London, which has been active right up to the present—even through the Apollo program, in which astronauts photographed the whole Earth.

For many generations after the Copernican Revolution new ideas were resisted actively by the Roman Catholic Church. (Only recently did the Vatican pardon Galileo for his heretical thoughts!) Analogously, today the open flow of information about extraterrestrials, free energy, and the Consciousness Revolution is running up against strong resistance among the contemporary establishment.

I believe the time we are in now will bring even more profound changes than those during the Renaissance. Copernicus, Galileo and others opened us to the reality of a three–dimensional, spherical world that turned out to be no longer at the center of the universe. The Consciousness Revolution we are now experiencing is opening us to new dimensions of reality beyond time and space. It also removes the human being from the top of the cosmic pecking order of sentient beings. This thought is so repugnant to some of those in power that it becomes ridiculed as being the stuff of tabloids.

The evidence is now clear to some of us that we are being visited by technologically superior beings whose agendas are complex and currently difficult to fathom. It is also apparent that our UFO visitors have mastered free energy and antigravity effects, as well as the ability to transfer abductees through walls and to pop in and out of our limited space–time dimensions.

In our arrogance we have woven our own parochial myths of contact [e.g., the ill–fated Search for Extraterrestrial Intelligence (SETI) program] within what we call the scientific community, and seem to be more interested in fiction than in fact, while the facts about extraterrestrial contact are obfuscated by those whom we think are the best and brightest among us. In my thirty–year history as a scientist, this has become a most curious phenomenon. The case of John Mack is a prime example of an old science which has gone into deep denial.

You may be skeptical of all this, which would not take away from the observed facts of free energy. Even the most traditional thinker who takes the time to look at the data will find that free energy is an acceptable concept. Such an investigator will also find that on the theoretical level we do not need to make any paradigm shifts in order to participate in the necessary research and development.

But as this explosive technology begins to penetrate the culture, our limited views will inevitably give way to entirely new perspectives. The paradigm shift is inevitable partly because the world needs it. We need a clean Earth instead of the one we are currently destroying. We have placed individual self-interest over the common good. My own challenge in all this is to maintain my scientific objectivity in the presence of powerful feelings, and to be able to forecast future consequences of the new world view. I believe that our transition into this new world will need to involve some deep grieving. But I foresee a happy resolution of all this as we become aware participants in the process of change.

"Apocalypse does not point to a fiery Armageddon," Joseph Campbell said, "but to the fact that our ignorance and complacence have come to an end." It is clear to me that, in order for us to survive and to thrive, we must now transcend that ignorance, lift the rampant suppression, and move ourselves into the new paradigm. Nobody will be immune from this process, and the grieving involved is a necessary and healthy step that will help lift us out of our resistance to change.

I know of few areas in which we need rapid social and technological change more urgently than in the use and abuse of our energy sources. Since the time of Nikola Tesla one century ago, we have seen fit to suppress those technologies required to clean most of the Earth's pollution, which in that same century has become rampant. It is apparent that we must soon shift away from our predominantly oil-based economy.

So this is a book about the personal and social impact of freeing this miracle from the void. It is also about freeing our own energies and going through the process of change. For if we are unwilling to do so our civilization faces a bleak future.

"UFO research is leading us kicking and screaming into the science of the twenty–first century."

–J. Allen Hynek, astronomer and UFO researcher

Meredith
Miller 95

2

How Our Fear Sets Up a Grieving Process

The real story of science and technology is not what I had learned about in school, seen in museums and taught in universities.

During my youth, the prospects of travel to the Moon and planets kept me glued to books on chemical propulsion and celestial mechanics. Having fulfilled my technical training, the next thing I knew, I was a Ph.D. and astronaut appointee on the ground floor of an exciting career in planetary science and astronautics. Soon I felt myself to be somehow "special," and my ego had grown to a point of seeming invincibility. I felt rather as if I were a young tenured Roman senator during the reign of Caligula or Nero, or perhaps the Catholic Church's most junior Cardinal during the Inquisition. As such, I would have been one of the least likely to point out that the emperor has no clothes.

In those innocent and naive times I was totally unaware of the suppressed mysteries of alien contact, UFOs, mystical experiences, spontaneous healing, near-death experiences and free energy. I had my elite status with all its financial and fraternal perks, to act as a buffer against any heresies. It was on this launch pad to fame and fortune that I was first exposed to the politics of power and greed.

Implicitly underlying all this education, recognition and potential was the rather grey, masculine world of the conquest and control of nature and the supremacy of the human over all other species. Something within me felt very unsettled about the bleak environments surrounding my hotshot astronaut and elite space scientist status, epitomized by the locker–room design of the space stations that are continually being designed and redesigned at the cost of billions of dollars to the taxpayer each year. Even then, I saw the politics of power play out from the inside, as the NASA culture rose to its prominence during the 1960s and subsequently fell into mediocrity during the 1970s, 1980s and 1990s. The death of NASA does not appear to have any signs of resurrecting.

And so I drifted from NASA center to NASA center, from prestigious university to prestigious university, looking for my answer, for the ideal nurturing environment for doing my science. But across the whole culture of science I found nothing that really resonated with a deeper essence. I had published several original scientific papers in my field, but somehow they seemed tailored to a mechanistic, hierarchal world view. Gone was my childlike sense of awe and wonder. I delivered physics gospel sometimes by rote to my students at Cornell, Berkeley, San Francisco State, Caltech, Hampshire, University of Massachusetts, and Princeton. I began to feel I had sold my soul and somehow knew it, yet had to keep up the charade.

The Energy Crisis of the 1970s

By the early 1970s I began to become acutely aware of an ominous trend in our culture. It seemed that almost everything done around me was motivated by personal gain, fear and greed rather than by altruism, whether it be the war in Vietnam, the proposed space shuttle, my own relationships with decision-makers within NASA and the universities—and the energy crisis.

In 1975 I became Morris Udall's science and energy advisor during his Presidential campaign and served as a special consultant for the U.S. House Interior Committee's Subcommittee on Energy and the Environment. Here I orchestrated hearings and wrote speeches for Udall on the deepening ecological crisis caused by our energy abuse. I was also unsettled by the dexadrine-popping, workaholic atmosphere created by ambitious advisors champing at the bit for their place in the White House and obsessed at finding the most "newsworthy" quotes for our candidate. What a superficial and contrived process newsmaking could be.

It was at this point, two decades ago, that I learned about the truth of what we were doing to the Earth and to ourselves. It had a lasting effect on me. Based on such scientific studies as the Club of Rome's report, *The Limits to Growth,* I had soon learned that the existing trillion-dollar-per-year global energy economy must convert from oil and nuclear power to renewable sources such as solar energy within a generation, or we would create irreversible pollution and would soon run out of Earth's oil. At the time, of course, I was unaware of the existence of free energy. I was also unaware that humans would later create a nightmare of unprecedented pollution and an economic system that was even more addicted to oil.

During the two decades that have passed since the Energy Crisis of the 1970s, annual worldwide oil consumption has more than *tripled* and electricity consumption has more than *doubled*. In my fifty-five years here we have used over half of the Earth's remaining oil in this most decadent of times. At our current gobbling rate, we have less than forty years of oil left. Here is a precious fossil fuel and petro-chemical that had been painstakingly formed by nature over eons and we have suddenly wrung out the lifeblood of the Earth in one greedy glut in our insignificantly short lifetimes!

Scientists have concluded that sixty–five million years ago, the dinosaurs probably became extinct because of the cataclysmic collision of an asteroid or comet with the Earth. Now they have literally become part of our fossil fuel base which we are ransacking as if there were no tomorrow. Caught in a pattern repeating itself, we are about to become extinct if we don't do something about it soon. Isn't it ironic that we are now in the midst of a cultural dinosaur fad? It is also curious that, as I write this, comet Shoemaker–Levy falls into Jupiter releasing energy the equivalent of millions of hydrogen bombs in its atmosphere—an Armageddon of similar magnitude to the dinosaur extinction event. Are these only mere symbols of our time or are they telling us something?

Since the time Ronald Reagan was elected President in 1980, the global energy–ecology challenge has been cleanly swept under the rug and it is now largely perceived in political circles as a nonissue. This is very sad. Besides modest reform efforts and the visions expressed by former President George Bush about a New World Order, whatever that means, we have nothing, nothing to seem to look forward to, while promising new opportunities fall by the wayside. For those of you interested in the public policy aspects of our history of energy abuse, I suggest you look at Appendix II (my paper "Green Power" presented to the Second International Symposium on New Energy in Denver, May 1994).

As a nation of sheep we seem to have accepted the oil spills, the nuclear proliferation, the Chernobyls, oil wars in Kuwait, more oil wars in Somalia, more drilling on the caribou wildlife sanctuaries of Alaska's North Slope just to buy maybe another year, the devastating landscapes of grid systems and refineries and tankers, the gas stations, the electromagnetic pollution, the foul air and waters, and the prospects of an irreversible global warming due to the overconsumption of coal and oil.

We are still burning fossil and nuclear fuels to boil water in large central power plants, a nineteenth century technology that is sapping our planet of its precious resources and ecology while we continue incinerating more oil in our (also outdated) internal combustion engines, clogging and choking our cities as if there were no tomorrow. And now China and other Asian nations want to expand that capacity. All of this is so unnecessary!

Why then do we continue to behave so destructively in the face of the obvious? How can virtually all of our leaders in government, in academe, in industry and in the mass media have forgotten such an important responsibility? My recent experience has revealed to me feelings of deep anger which I am freely expressing.

The Politics and Science of Fear

I suggest that there is a deep-seated, mostly unconscious fear within the governing body and within most of us that our entire economic and political system would have to fall were the necessary steps for change to be taken. Our polluting global energy infrastructure now consumes about a fourth of the Gross World Product and an even greater share of what is controllable by powerful vested interests. Other aspects of world economies, such as delivering food, clothing and shelter, are more decentralized and cannot be manipulated as much.

My efforts to help unsuccessful Presidential candidates George McGovern and Jesse Jackson on their economic conversion programs were modest in comparison to what is implied here. History shows that such efforts have no political capital even though they would have retrained aerospace workers from producing weapons to doing activities which would clean up our planet. It seems that fear, greed and shortsightedness are on winning political sides. The only choice of candidates these days seems to be between Tweedly-dee and Tweedly-dum. In staying home, most voters seem to be saying "none of the above."

Fear and Greed in Other Choices in Science, Technology and Medicine

How we use and abuse our energy sources is but one example of our cultural illness. The underlying fear and greed can be seen across the board. For example, Tulane University philosophy professor Michael Zimmerman gives cogent reasons why the establishment elites resist the very idea of encounters with aliens with superior technology. In these cases we find very unscientific approaches to the truth among our leading scientists. In his zeal to debunk the evidence, astronomer Carl Sagan seemed to go to extremes in distorting existing data on the "face" on Mars and about the abduction phenomenon in popular articles for Parade Magazine. Sagan sees abductions as hallucinations, says there is no physical evidence of UFO phenomena, and seemed to contrive a second photograph of the Mars face which appeared to be doctored from the original version to not look like a face. Either Sagan is totally unaware of the available data or has become a disinformation specialist for the existing world view. More on this later.

Either way, while many unsuspecting observers think of Sagan as the guru for alien contact, in a new paradigm setting he would become a mere storyteller of a long lost myth that favors one particular "safe" form of contact (the SETI program of listening to hypothetical radio telescopic signals sent from more than tens of light years away). Yet he dismisses the reality of contacts that are clearly showing results, even though the findings and the methods of investigation appear to be unorthodox to an unsuspecting astronomer. In avoiding the obvious evidence, Sagan perhaps represents the very archetype of a member of the establishment elite who harbors such deep fears of the aliens' agendas which do not match his; he sees fit to present himself as a scientist, yet he seems to demonstrate just the opposite! Ironically, Sagan is an inspiring leader of the global ecology and weapons reductions programs.

It seems nowadays that one's investment in a particular world view is more important than the truth. Even our most respected scientists have become propagandists in the name of science. These individuals then become like the Cardinals of the Inquisition. The methods of the Inquisition seem to be with us, except now they are cleverly disguised. I am sure Carl Sagan does not wish to go down in history as an apologist for a destructive status quo, but he may not realize what he is doing.

Similar stories are told about the medical profession's resistance to alternative healing. Across the board, we are seeing at least two truths emerge: the official "truth" as told by those on the government dole, and the emerging unofficial truth based on rapidly mounting scientific evidence and extensive personal experience and reported in the alternative press. The gap now seems to be broadening between the two perceptions of truth, and this can lead to a great deal of anxiety for seekers of the truth.

Political Expediency and Hidden Agendas

I see the gulf between the two realities as being so far apart now that members of the ruling class have to resort to extreme measures to suppress reality, just as a church cleric would have done during the early Renaissance. The tech–

niques include excessive secrecy, outmoded patent and securities laws, media blackouts on new developments, assassinations, suicides, threats, illegal domestic espionage, propaganda, ridicule, debunkery and lack of financial support, just to name some.

All told, we are in a time of paradigm shift, as suggested in the authoritative text *The Structure of Scientific Revolutions* by Thomas Kuhn. According to Kuhn, mainstream scientists' denial of obvious new data is a telltale clue to a paradigm shift, or a totally new perspective on reality.

In his thought-provoking *The Seat of the Soul* author Gary Zukav sees the old paradigm as being motivated in part by the seeking of external power (position, prestige, money) and the new paradigm as the use of authentic power (sourced from inside and with compassion for others). As we shall see later, both Zukav and I feel what is coming is even beyond paradigms. We are on the threshold of a great evolutionary leap that transcends the intellect.

Masquerading the Consciousness Revolution is the ivory tower of babble of old "truths" proliferating in this Information Age. If an outmoded and polluting Industrial Age is to continue to dominate the world through the deeply unconscious forces of fear and greed, the Information Age is merely a mouthpiece for a decadent world view. Most of the mass media reinforce this grievous distortion, and we become mesmerized by the barrage of disasters we seem to be able to do little about and pseudo-truths we hear on our daily television and newspaper reports.

But the sheer increased volume of available information does not guarantee the rapid dissemination of the truth. *Garbage in, garbage out* pertains to any form of information, regardless of how many megabytes or how sophisticated its form may take. The irony is, in the presence of the new paradigm which those of us in control especially find so fearful to address, the Information Age at levels of deeper meaning becomes the Disinformation Age. Information is only as good as its quality, but I find in what is called "news" is mostly information at a superficial level and often disinformation at the essence level. Memories of Orwell's 1984.

So it is no wonder strange things are happening. It is no wonder so many of us are ignorant and complacent. The coming changes are so vast, nobody in power wants to acknowledge it, manage it, or participate in it, except perhaps in secret. Many of us seem to trust the media or leadership or somebody in charge to make the dramatic announcements heralding big change. But nobody seems to dare do it for fear of being rejected, ridiculed, persecuted, fired...or freed.

Freed? Yes, that is also part of moving through the fear. What would we do if we were not doing all the safe and familiar things we do now? Unconsciously we seem to be afraid of new possibilities. So like sheep in denial grazing under polluting power lines, we wait for the other shoe to drop.

A Washington Post reporter who gave me an in-depth interview about free energy last year ended up publishing a safe historical curiosity piece on Tesla, with no mention of contemporary free energy. Did his original piece get axed by his editor? Did the science editor tell him Carl Sagan wouldn't approve, or was this new blackout from a higher place? Do the aliens have the Government under gunpoint not to reveal the truths of their presence and the existence of free energy?

Whatever, those in power are deeply afraid of the coming changes. Perhaps I would be too if I were still part of the establishment elite. If politicians had even been afraid to implement modest economic conversion programs for aerospace workers, how could anyone in power and connections with existing infrastructure possibly sanction freedom from oil interests and the grid system?

Fear, Greed and Grieving

Many of us living in the United States, have a deep–seated fear of the unknown, in part because of our comfort and supremacy for at least fifty years. We have become hooked on money as our idol. We have used the acquisition of money as our hidden agenda for raping the Earth. Powerful entrenched interests, secrecy and money have become the tools to reinforce the old orthodoxy in our times. This has led to widespread greed, the kind that sends our captive leaders into undercover, corruptive, unethical and illegal behavior.

Obsessed by control and self–protection, we have distorted the truth and moved sideways into a land of daily pseudo–news. Meanwhile the real news, for example the development of free energy, has been totally emasculated by reporting on putting out brushfires. We have become distracted crisis managers, assuming the world will always be the same as it is now, save for new wars, disasters, murders and corruptions.

The world has no choice but to change in spite of it all. Meanwhile we appear to be hanging on for dear life, living a myth we ultimately know is wrong for our children and for their children. A basic hypothesis of this book is, we will know nothing about bringing in the new paradigm until we are willing to grieve the old.

This is largely an emotional process, not an intellectual exercise. The intellect may help in pinpointing an obvious problem and its possible solutions. But it can also become a trickster directing us into denial or bargaining, keeping us from experiencing the emotions of grieving we will need to do to bring in the new.

I am personally in the midst of a grieving process which is deeper than I had bargained for. The coming changes will require that all of us express feelings about them. For those of us bold enough to pioneer this new territory and to accept the new ideas, we will need to experience both the losses and gains of the shifts. Barring miracles (which I am learning *are* possible), the rest of us will need to go along, even if kicking and screaming, because we have a fragile, dying planet to help transform. Winter has come and we may not make it to spring.

Because the implications of the paradigm shift will be so radical and mostly unpredictable even by a futurist such as myself, the resistance to the truth will probably continue to mount until we can all face up to the truth. Grieving can then become a conscious process.

In contrast to the emotions, the intellect can provide a more convenient way to visualize the enormous implications of shifting our priorities from selfish exploitation to cleaning the Earth, discovering our true nature and joining the Galactic Club of sentient intelligent beings. Some of us may not like how all that is unfolding, but as the public revelations about these inquiries become better known, the shifts will become apparent to all of us.

According to the research of Kuhn and others, during the Copernican Revolution and during other significant changes in history, the establishment elite stuck to their beliefs and held on to their power until they died off. At present, we don't seem to have time to do all that and to preserve the Earth—hence the need for many of us schooled in the old ways to open ourselves both intellectually and emotionally. We must begin the work of grieving our outworn patterns. I believe that only then can we begin to make a difference during the coming global transformations.

The concept that grieving is a healthy and necessary process is not a new one. Psychiatrist Elisabeth Kubler-Ross and others have researched the emotions which psychologically healthy people go through when they are given a diagnosis of an "incurable" disease or when they lose a loved one. She was able to isolate a sequence of (sometimes overlapping or retrospective) feelings and behaviors most people need to experience before they can accept the death. This landmark research showed that many of us have difficulty acknowledging or expressing these feelings and so may be stuck at a particular stage for as long as years or a lifetime of unhappiness. The stages include denial, anger, bargaining, depression, and acceptance.

"The more we are making advancements in science," wrote Kubler–Ross, "the more we seem to fear and deny the reality of death. How is this possible? We use euphemisms, we make the dead look as if they were asleep, we ship the children off to protect them from the anxiety and turmoil around the house if the patient is fortunate enough to die at home, we don't allow children to visit their dying parents in the hospitals, we have long and controversial discussions about whether patients should be told the truth..."

In other words, the prospects of our own inevitable physical death or that of a loved one is riddled with denial. Culturally, we become like the lost children described by Elisabeth Kubler–Ross. The parents play the role of auth–ority, the purveyors of secrets. Only by addressing the real issues and expressing our grief will we get beyond our denial.

I am suggesting we will all need to grieve the old cultural paradigm before we can embrace the new. At this moment, the vast majority of us are trapped in a labyrinth of denial about the reality of the impending death of our old awarenesses.

Some may argue that the death of the old paradigm is not imminent. In the absence of momentous Earth changes or wars it sometimes feels quiet on the home front. But that can be deceiving. Earth is like a terminal patient who is given the diagnosis of death before having felt the full brunt of her disease.

The inevitable future points toward radical change. New thoughts, experiments and intentions are being born, waiting in the wings for their opportunities. At a sub–conscious level, I believe the death of the old paradigm has already happened and we are in a mass denial similar to that of Galileo's colleagues who refused to look through his telescope because they didn't believe it could make things look bigger. A poignant contemporary example is the Harvard faculty committee who refuses to see any merit to John Mack's work on UFO abductions.

The more steeped we are in the old ways, the deeper will be the grief. Some of us have lost a spouse while others may have only lost a cousin. A not yet recognized truth is we are all losing our Mother Earth so we will all need to grieve the old, some of us more than others. The higher we climb, the farther we may have to fall. Having been an old–paradigm scientist for most of my life, I am finding I have to grieve a lot. All this seems to be a normal response to the changes we must face.

We are all familiar with our initial feelings of loss. To defend our former belief (e.g., the doctor's diagnosis must be wrong), we erect armor to deny the announced reality. It didn't happen! It can't be true! I didn't hear that! Don't tell anybody! Shoot the messenger!

We have become a culture enmeshed in denial and suppression. In the next chapter we will look at how our collective denial has blocked the application of free energy from the time of its birth 100 years ago with the remarkable developments of Nikola Tesla, right up to this very moment: Now dozens of Teslas are proving the concept, yet hardly a penny from the billions of dollars spent each year by the U.S. Department of Energy goes to free energy research and development.

900

3

Denial and Suppression

"The cruelest lies are often told in silence."
–Robert Louis Stevenson

"The purpose of the international conspiracy is to maintain a workable stability among the nations of the world and for them, in turn, to retain institutional control over their respective populations. Thus, for these governments to admit that there may be ... technological capabilities obviously far superior ... could, once fully perceived by the average person, erode the foundations of earth's traditional power structure. Political and legal systems, religious, economic and social institutions could all soon become meaningless in the mind of the public. The national oligarchical establishments, even civilisation as we know it, would collapse into anarchy. Such extreme conclusions are not necessarily valid, but they probably accurately reflect the fears of the 'ruling classes' of the major nations, whose leaders (particularly in the intelligence business) have always advocated excessive governmental secrecy as being necessary to 'preserve national security'."
–Victor Marchetti, former special assistant to the Director of the CIA, quoted by A. Michrowski

In *The Second Coming of Science* I conceptualize reality as a patchwork quilt of boxes, each box representing a belief system. The predominant belief system of our time is Western science, which tends to look at everything in objective, concrete, materialistic, reductionistic and deterministic ways. I have mentioned that the ultimate expression of such a world view is the hypothesis, advanced by astronomers Stephen Hawking, Carl Sagan and others, that the mind of God could be known if only we reconcile the known forces of physics (gravity, electromagnetics, strong nuclear and weak atomic forces) into a physically limited unified field theory, one that could track reality deterministically from the moment of a cosmic Big Bang about 15 billion years ago.

In such a decadent view, no room is left for the creative actions of consciousness. And yet, in my journeys around the world, I have found untold violations of this basic paradigm of materialism. Aspects of the new physics, free energy, alternative healing, contact with the dead and with nonhuman intelligence, UFO activities, and psychokinesis are phenomena that clearly point the way to a new science, based on innovative exploration and on observing the interactions of consciousness with the material world.

My main focus over the past ten years has been a scientist's quest for the truth. I have examined those theories, experiments and demonstrations of reality that make it clear to me that the box of Western science limits us severely in both time and space coordinates. On the time axis we look only at events between our birth and death, while ignoring near-death experiences as well as evidence for reincarnation, spirit communication and spirit possession. On the space axis, we exclude our inner and outer experiences which reflect "anomalies" such as psychic and UFO phenomena whose importance is relegated to unsupported "pseudoscience" and the tabloids. Most of those who consider themselves as scientists still deny the truths outside the box.

The tenets of Western science have led to industrial technologies that have quickened human–caused massive deterioration of the environment. We seem to have little time left to make the needed changes, a generation at most in my opinion. For example, we are still burning what little oil is left in the world at an unprecedented rate that will deplete us around 2030. We are destroying our atmosphere with pollution from obsolete internal combustion engines and power plants.

We shall later see that a moderately funded public or private research and development effort to make free energy available to the public is almost certain to bear fruit. As a global culture we seem to have the technologies required to transform the planet and to provide unprecedented growth in human opportunity. The contributions already made by the pioneers of free energy, alternative medicine and other promising new sciences make it clear that we now have the ability to transform the Earth into a Garden of Eden, free of the blights of overindustrialization. Of course, we also need to solve the problems of overpopulation. In principle, most of our individual needs can be met simply, inexpensively and cleanly.

So why haven't we done anything about this? Gradually over the years, as I looked at one case after another, I discovered that we have created an environment of insidious suppression of the most promising new technologies. We have created, promoted and allowed the suppression to so permeate our minds and hearts, we have lulled ourselves into a dangerous ignorance and complacence.

Underlying our head–in–the–sand behavior are economic systems that are hooked on the growth of existing, outmoded systems that are ruining the environment. We are so heavily vested in these systems, we have become imprisoned by borrowing from an impossible future that we have economically preordained. We seem to be like children locked in an electric jail. We have given away the keys to faceless technocratic captors and have made the economy our God.

In many ways, the environment in the United States for suppressing important new inventions seems even greater now than it was a century ago. "It was possible," wrote Canadian political scientist Andrew Michrowski, "for Nikola Tesla, Alexander Graham Bell, George Westinghouse and other great introducers of beneficial technology to make their mark because in their times, before 1913, the retardant forces were not yet organized enough to totally counterweigh these innovations, although these forces were devious enough to twist the intent towards such politically rewarding aims as warfare, massive capital-intensive enterprises and the regular scheming, wheeling-dealing exercises so favoured by the successful lot."

Also underlying the juggernaut of misplaced resources is a world view, particularly in the United States, which reflects what philosopher Michael Zimmerman calls "anthropocentric patriarchal humanism." This is a sense of macho self-determination, autonomy and superiority over all other living things. Under this paradigm, man feels he needs to be supreme and free, whereas nature is viewed as raw material to be exploited. There is no room in this belief system for the very idea of nonhuman intelligence, let alone free energy and alternative healing. Backed by various fundamentalist and other conservative religious belief systems, as well as by leading scientists, technocrats and media, this outmoded philosophy plays itself out in a military-industrial-financial complex that is literally destroying all of us if left unchecked.

Leading ecologists have often stated we are not only inheriting this Earth from our parents, we are borrowing it from our children and grandchildren—so much so there may be little left for them. Our denial of this fact has kept us from proceeding toward solutions. The deteriorating relationship among the three E's, environment, energy and the economy, contains the core of our challenge.

"Energy is such a vital aspect of our civilization," wrote Michrowski, "that its misapplication has been bringing down governmental function to near-operational levels and fiscal bankruptcy since the 1960s and if governments as we know and cherish them are to survive, they must grapple with Energy with the same determination as National Survival."

I believe these scientific, economic and philosophical systems must be transformed at their root level if we have any chance to save ourselves from global catastrophe. "No problem can be solved from the same consciousness that created it," said Albert Einstein.

And yet those prevailing beliefs so permeate nearly all of us that we all set a stage for the suppression of any and all discoveries that could lift us out of our own destruction.

Such is the nature of a paradigm shift. During the Copernican Revolution, the world grew to know that the Earth was not flat, nor was it at the center of everything. During the Consciousness Revolution we are now in, we are learning that we humans are not at the top of a cosmic hierarchy and that materialism is but a subsidiary aspect of our reality. Those who call themselves scientists seem to be among the last to make the shifts. Like the church officials of the early Renaissance, these "keepers of the truth" have seen fit to deny and suppress new evidence that threatens their own coveted world views.

And so, in the presence of overwhelming evidence, the box of Western science–economics–philosophy is ready to burst open, rendering the old view as obsolete. It is at these pivotal times that orthodoxy will do all it can to keep the box sealed off as tightly as possible, like a parent denying their child the truth of the impending death of a family member. While its cracks are readily discernible by those intrepid individuals who dare to step outside it, the awesome power of vested interests is doing everything, *everything,* to plug the leaky dike. These interests are often oblivious to the possibilities that lie ahead. Here I am not talking about New Age claims or pseudoscience that have little independent verification. I am talking about solid scientific evidence that is pervasive and available to those who choose to look.

To further develop the analogy of the denial of an imminent or recent death in the family, many of these suppressive activities arise from subconscious fears and may not be premeditated or conspiratorial in the classic sense. "When a patient is severely ill," wrote Elisabeth Kubler-Ross, "he is often treated like a person with no right to an opinion. It is often someone else who makes the decision if and when and where a patient should be hospitalized."

As a culture, we too are being secretly guided by others, perhaps because we are unwilling or unable to face our own truths. It seems that parts of the intelligence community, true to their calling as being intelligent (or at least more so than others nipping on their heels), have capitalized on this and have played the role of our parent. Perhaps this unwitting alliance between a CIA and a Sagan can keep the dikes plugged awhile longer. For the scientists' bias against the evidence, in the name of the skeptics' adage, "extraordinary claims require extraordinary proof," comprise a leading public force that continues to keep us from the truth.

This bias toward denial therefore supports the creation of an elite corps of segmented and loyal suppression agents to confront the truth for us. Eyewitnesses to extraordinary events such as the well-documented 1947 Roswell UFO crash are beginning to feel free to come forward only decades later—bordering on safe historical distance. Likewise the suppression of Tesla's work is now more freely discussed, but in a historical rather than a contemporary context.

Could it be our own collective unconscious fear of the Pandora's Box about to open that drives the rampant suppression of promising new technologies? Is that fear keeping many competent scientists away from investigating these things in detail because of peer pressure and funding restrictions? My experience with this gives a strong "yes" to these questions. I have earlier suggested that because the implications of the Consciousness Revolution exceed even those of the Copernican Revolution, the resistance and denial is far greater now than then. It could be staggering if we follow Bertrand Russell's axiom that the resistance to a new idea is proportional to the square of its importance.

The Suppression of Free Energy

In 1993 I co–founded the Institute for New Energy which has so far held two "think–tank" retreats and public symposia in Colorado. Both meetings included scientific papers which present ample evidence for the theoretical and experimental efficacy of free energy, which has the potential to replace traditional energy in the near future. So far nobody has succeeded in making these devices available for individual use, but I believe we are close. We are still in the research phase of a research and development effort. Commercial models most likely to be made in Japan will probably become available within the next few years.

The governments and private industry in India and Japan are funding top level scientists and engineers to develop free energy for commercial applications, something about which the American government appears to know little or nothing. Cold fusion pioneers Martin Fleischmann and Stanley Pons, formerly of the University of Utah, are now in France being funded for their work by a Japanese consortium. The inventor of the N–machine Bruce DePalma, formerly of

MIT, is now developing his free energy concepts in New Zealand. Other American inventors and researchers have gone underground most of the time (e.g., Thomas Bearden and Sparky Sweet), have been sued (Sweet), had their devices taken away by impatient investors with gold fever (DePalma), had their machines confiscated by the Government (e.g., the Canadian inventor John Hutchison and American Dennis Lee), been convicted and jailed under questionable charges (Lee) and in at least one case have been told by the Government to change careers—or else (e.g., Adam Trombly).

In all, I have met several dozen free energy researchers. What all these individuals have in common is the underfunding of their work such that it proceeds to proof-of-concept but no further. Developing useful prototypes requires a much larger effort as would come from bringing the researchers together in a research and development effort analogous to the Apollo or Manhattan projects. But there has been no public and little private support for free energy inventors—particularly in the United States—even though this country is where most of the ideas come from. We seem to be so active in suppressing this technology we have driven most of our brightest inventors away or underground.

The remarkable fact is, we seem to have had this technology for one century! Nikola Tesla was among the first of a band of energy mavericks, who through the decades, have repeatedly demonstrated free energy, only to be suppressed later. For a whole century we probably didn't have to pollute the Earth to meet our energy needs!

Sometimes matters turn out to be that simple, and we scientists will need to open our blinders, to step outside the box. We can then confront the obvious while admitting to our former ignorance. In the case of free energy, however, there is more to the suppression than the mere denial of leading scientists and engineers. History suggests that J.P. Morgan withdrew funds from Tesla, perhaps because he discovered Tesla's energy device might threaten Morgan's potential control of the new power grid system which would use wires from Morgan's monopoly on copper mines.

Since the time of Tesla, free energy has been 100 per cent dammed up—a remarkable achievement for the suppressing forces— for any one or more reasons: harassment or death threats from government or energy industry agents, lack of adequate financial or engineering input, personality conflicts, and little support from the technical community, the press and from the entire world.

But perhaps the largest obstacle for inventors in the United States is the threat of being victim of the Gag Rule, or Secrecy Act, which the Government has already placed on some free energy inventions such as those of Adam Trombly. The Government confiscates your invention and your work goes down the drain; it's as simple as that. Therefore most free energy devices are not patented until they are close to commercialization, which is a big step most inventors never get that close to, as product development is usually outside their expertise.

Gag rules and isolation are not the only problems experienced by the inventors. Bruce DePalma seems to have had no end of difficulty in dealing with his own investors. In recent months, he has had his N-machine prototype taken away from him because of deep philosophical differences. His investors wanted to have a timely return on their financial commitments. As a former MIT physics faculty member, DePalma wanted to tinker with, improve, and understand the basic science of free energy production. The confiscation happened only after some of the investors were becoming impatient. To justify their actions, they ended up hiring a debunker to superficially analyze and discredit DePalma's work. Investors often don't seem to understand the basic importance of this kind of work to the world. So far no investor in this field is yet receiving monies from free energy technology because we are still in the research cycle of research and development (R&D). When we do get into development, the situation will change quickly and dramatically. Meanwhile a kind of gold fever seems to be dominant.

Many other American inventors have spent years seemingly suppressing their own potential patents while they patiently wait for the necessary breakthroughs to protect their own interests. They are playing the patent game by the book. Alongside a handful of inventors, each of these groups are isolated, anonymous, and await commercial opportunities which never seem to come. During my visit to one such inventor, a commercial prototype was awaiting detailed testing by a part-time electrical engineer moonlighting at lunchtime from his "real" job. This is hardly the stuff of a team preparing for a moonlanding or a hot fusion project. I am in touch with other isolated inventors who have had similar difficulties and who do not want to be publicly mentioned at this time because they fear suppression.

British inventor-physicist Harold Aspden lamented the problem this way: "There is no way forward for anyone involved in real research on free energy from ferromagnetism, unless that person understands the physics of the subject. The hit-and-miss ventures of those who build permanent magnet 'free energy' machines and get them to work anomalously only guides others equipped with the right training to take the research forward."

In other words we seem to need a team of qualified inventors, scientists and engineers to bridge the gap from proof-of-concept to implementation. This lack of organization has been potent enough to keep free energy under wraps for 100 years, and yet it seems amazing that, given our extraordinary ability to investigate and communicate about new concepts in science and technology, we have overlooked the potential of such a huge industry, whose output is equivalent to unimaginable trillions of dollars now spent on environmentally abusive technologies. The new industry would manufacture billions of free energy devices at far lower economic and environmental costs. Even if you do not accept my informed optimism about this, it would seem wise for someone to spend a few million dollars now to take a look at what will almost certainly be the largest force to clean up the Earth.

Could it be that such airtight suppression comes from a collective denial of the death of traditional energy, that somehow we don't want free energy to happen? In spite of all that, we appear to be very close to building commercial units, and the knowledge is dispersed among many technical people worldwide. I believe the genie of free energy is now out of the bottle. No amount of suppression will be able to stop this one.

On the broader question of the suppression of new science in general, preliminary results of my research show that the external sources of suppression come from different communities, according to the general category of investigation. For example, in the free energy area, we see the external forces of suppression as being a blend of factors that inhibited things enough to ever reach a critical mass of implementation for 100 years.

On the other hand, the suppression of UFO and other extraterrestrial intelligence information for at least 48 years is probably being orchestrated by an elite band of men in the CIA, National Security Agency, Defense Intelligence Agency and their like. This small group appears able to keep these already-hard-to-believe secrets very well.

The Mars Anomalies: A Case Study of Scientific Denial and *Ad Hoc* Pronouncements about a Reality

Having spent approximately twenty years of my career as a full–time planetary scientist at various universities, I became a specialist in the study of Mars. During some of that time I have served NASA as a scientist–astronaut potentially destined to go to Mars, completed a Ph.D. thesis on the optical properties of the red planet and published dozens of peer–reviewed research papers about Mars, Venus and Mercury. I also gained experience as an imaging scientist and was Deputy Team Leader on the Mariner 10 Venus–Mercury Television Science Team.

In the early 1970s I spent some years during the Mariner 10 mission at the Jet Propulsion Laboratory and Caltech on the scientific planning, operation and interpretation of spacecraft–taken images of planetary surfaces. Michael Malin, chief scientist of the recently failed Mars Observer Camera, had worked with our team at that time as a Caltech graduate student.

Over the past decade I have taken an interest in the "face" and other anomalous features on Mars photographed in 1976 by the unmanned NASA Viking Orbiter and described in my earlier books. Besides reviewing other researchers' work, I published a peer–reviewed paper on an analysis of the "face" showing that its three–dimensional structure is facelike and not a trick of lighting and shadow—contradictory to the conclusions of my mainstream planetary scientist colleagues. The independent work a number of us outside of NASA have carried out reveals strong, but not yet conclusive, evidence for the presence of artifacts of an intelligent civilization on Mars.

The ensuing controversy between the NASA–funded scientists and several qualified investigators outside the system—and NASA's refusal to accept the possibility that an extraterrestrial artifacts hypothesis could be entertained—has been well-documented by Sonoma State University philosopher Stanley McDaniel in his recent book *The McDaniel Report.* He cites many instances of neglect and obfuscation by NASA, Malin, astronomer Carl Sagan and others. A number of us have concluded that the extraterrestrial artifacts hypothesis has considerable merit, based on state-of-the-art image processing of the Viking photographs.

Unfortunately, the signal from Mars Observer, which could have sent us the needed high resolution photographs of the face and surroundings, mysteriously ceased just as the spacecraft approached Mars in August 1993. Several causative hypotheses are possible: (1) the loss of signal was due to a spacecraft or command malfunction, (2) the commonality of this situation with the Soviet Phobos 2 probe suggests some form of intelligence interfered (either extraterrestrial or maybe even from a NASA negative group mind), or (3) NASA has decided to conduct the mission clandestinely by faking the loss of signal, so pictures could be kept from the public, pursuant to an established policy of secrecy. In his book, McDaniel cites the 1961 Brookings Institution report, which advocates secrecy in dealing with discoveries of artifacts on other planets because of culture shock and threats to the careers of government officials and the scientists themselves! Is it conceivable that such an anthropocentric view could justify a decades-long policy of official denial and suppression of evidence? Perhaps so.

It is clear to me that a large segment of the American public would like to have the face and surrounding area be re-imaged by a camera at higher resolution and at varying sun angles as a target of opportunity. We could carry out such a mission at low cost using a Pentagon spacecraft called the

Clementine, but political obstacles are formidable, not the least of which is our current policy of secrecy.

The Search for Extraterrestrial Intelligence (SETI) is a compelling inquiry of great philosophical interest to all humankind. But the scientific mainstream and NASA have seen fit to limit its own SETI charter to only one contact scenario that is unlikely to succeed because of an assumption that extraterrestrials would wish to match a particular technology which is insignificant on a cosmic scale: radio astronomy. Yet, in spite of its relatively credible stance with the scientific mainstream, Congress recently canceled that program too. Meanwhile, some very intriguing evidence on Mars may be staring us in the face. We also have abundant evidence from UFO sightings, abduction reports, crop circles, physical ground traces, cattle mutilations, etc., ignored by NASA and its scientists.

An emerging new paradigm challenges the widely-held view that scientists can be experts on just about everything. The new paradigm admits the possibility that any evidence of intelligent extraterrestrial activity needs to be thoroughly checked out. The scientists' denial is largely unacknowledged by our mass culture.

Whether by bureaucratic incompetency, arrogance or conspiracy, NASA's record on the Mars face inquiry shows the same kind of closed-mindedness. In these late stages of preparing for relaunching a mission, it would seem prudent for us to transcend these limitations and come together as brothers and sisters in true cosmic exploration and enlightenment. But if a secret NASA agenda were to be revealed, I would suggest that most of those individuals who are responsible not be prosecuted except in those cases where individual civil rights have been clearly violated. The same reasoning would apply to UFO and free energy research. Legal vindictiveness, in my opinion, has no place during such a radical paradigm shift. Leaders of the new paradigm need to be forgiving and not repeat the mistakes of the past.

The interesting case of the Mars anomalies coverup seems to stem from a curious partnership between mainstream scientists and NASA, temporarily in alliance to keep their own projects and other vested interests intact. The suppression here is by the Government and by mainstream scientists who depend on NASA for their funding. Both groups are the only ones with frequent access to the media, and the truth becomes compromised for convenience. We are a culture in denial of new truths.

Suppressions in Parapsychology and New Medicine

The suppressions in parapsychology appear to come from the harangues of the skeptics' groups and the parapsychologists themselves, many of whom are overly conservative and narrowly focused. In one such case, a religious group who you would think should support scientific evidence for the healing power of the mind has done just the opposite. Christian Science practitioners Bruce and John Klingbiel were rejected by their own church for most of the past eighteen years, during which they had done excellent research backing the tenets of the church on the efficacy of prayer. It seems the church hierarchy felt they needed no more proof of the power of the mind, and so turned against efforts to add more evidence for what they felt they already knew to be true.

The parapsychologists also dismissed the Klingbiels' work because the language used in their research was not befitting that of a Ph.D. psychologist. Then, when one parapsychologist, Dean Radin, did replicate the Klingbiels' important work on amplifying psi effects in concealed card guessing, their results gained new credibility. The story of the Klingbiels came to a sad ending when this downtrodden father-and-son team committed an apparent suicide last year. They were modern martyrs caught in the crossfires of the depression phase of grieving.

As for suppression in the healing and health arena, the main perpetrators include the doctors themselves, who adhere to the comfortable lifestyles of their specialties, which in concert with the drug companies, healthcare equipment manufacturers and hospitals, provide a powerful vested interest in drugs and surgery as the primary methods of treatment. In the medical establishment, consideration is rarely given to the healing power of the mind and most other holistic approaches. Even most university scientists, who are the ones we might believe keep track of developments at the frontiers of their own fields, give no credence to these possibilities. The larger community of biologists and chemists have ignored the concepts of the new physics and consciousness, and are still stuck in the old paradigm of materialism.

A notable example of this form of suppression was the National Science Foundation panel on parapsychology, that rejected the solid research by polygraph scientist Cleve Backster on biocommunications (telepathy) because of their own prejudices about language and experimental approach. Yet his very important positive results have been shown to be borne out in similar experiments by others. Backster's lack of funding and of interested colleagues keeps his experiments at a rudimentary level.

And, of course, we hear about miraculous cures for cancer and AIDS, which could have come from the microscopes of Raymond Rife and of Gilbert Naessens, both well-documented suppression cases (see the work of Christopher Bird in *Suppressed Inventions* edited by Jonathan Eisen). In both cases the inventors have had no end of trouble with the medical, research and governmental systems.

In most of these cases of suppression, outside investigators have had years of experience, as opposed to the kind of *ad hoc* or *a priori* thinking mainstream scientists often use when debunking new science. To the traditionalists, any bold hypothesis is unlikely to hold, and therefore we don't need to use any science to debunk it. It seems that the more time I take to sift through the evidence the more I am amazed to find that we have unwittingly and collectively created the myth that reality is very limited. Nobody seems to want to take responsibility for the suppression. Yet we seem to be ready now to transcend the bonds of limitation to the truth about the myriads of suppressions we have all been perpetrating.

The Pervasiveness of Suppression

When we hear about suppression in Government, many of us think of isolated incidents such as Watergate or Irangate or the development of a particular Star Wars technology. We don't tend to think about the suppression as being so deep, so pervasive. Perhaps the reason we haven't had the benefit of knowing about the extent of the suppression is that we were all in denial. From a broader historical perspective, it may seem odd that we must walk that more difficult path, before the truth can become available to us.

In other words, we may not have been ready for the truth.

It is difficult for those of us who claim to be scientists to have to admit we have our own gaping blind spots. Nobody wants to take responsibility for his own oversights. No formally organized group of, say, a thousand or more individuals has ever gathered long enough to carry forward any of these gauntlets of innovation and social change, any one of which can profoundly change the world. It is no wonder we have been in the dark. But the time is nigh to admit it and to carry on with what needs to be done.

It appears that those scientists, industrialists, journalists, and governmental officials who have perpetuated the suppression are now desperately holding on to a myth that no longer works. It is normal to get angry at the deceivers. I suggest we forgive them and offer amnesty to all but the most invasive ones, so that we can all get on with the new paradigm! We will also need to offer our colleagues support through their own unsettling losses of former positions so that they too can take part in our exciting new adventure if they choose to.

Ours is a century of suppression of alternatives to our polluting ways. One hundred years ago the sexual suppression of the Victorian period began to lift. These forces may have transferred to our obsession with secrecy in the presence of potential wars of massive destruction. But what was also sacrificed was finding out what is truly possible, thus sabotaging our potential for a clean and uplifted world. But there are leaks at the edges, tattered cloth at the table of Gaia, perforations that are opening and gaping and transcending the suppression, I think once and for all. The Suppression Syndrome is at last coming to light for what it is. Understanding denial and suppression is a mixed blessing, a humbling experience on the way to enlightenment.

Reviewing the Sources of Suppression

In my studies of the Suppression Syndrome of our times, I have been able to identify the following sources of suppression of important new developments in science and technology.

* **The scientists themselves**. Throughout history the curious phenomenon of scientific denial is what prevented Galileo's colleagues from looking through his telescope, or what caused the French Academy of Sciences to deny the

existence of meteorites, what makes contemporary astronomers ignore the UFO or Mars Face evidence, or what makes Harvard University want to appoint a secret committee to fire John Mack. In his classic text *The Structure of Scientific Revolutions*, Thomas Kuhn has clearly shown that scientists can be very unscientific when their own world views are threatened.

* **Industrial suppression**. Oil and utility executives generally do not want free energy to happen. General Motors does not want electric mass transit. Pharmaceutical companies and doctors do not want to see their profits end with a miracle cure. Banks and insurance companies don't want to handle any big change. In each case, a multibillion dollar vested interest perceives a threat to its own wellbeing, spawning a governmental–industrial alliance of powerful special interests.

* **Governmental secrecy**. This alphabet soup of agencies (CIA, NSA, DIA, Office of Naval Intelligence, etc., etc.), born of World War II and superpower confrontations, spends $35 billion or more each year of the American taxpayers' money to sustain an unaccountable suppressive force that is now abusing its power. The Iran–Contra deal was but a tip of an enormous black budget nightmare of an iceberg. Because it is sometimes impossible to distinguish information from disinformation, and because secrets are compartmentalized and based on need–to–know, those who have investigated this hydra–headed beast believe that the Cosmic Watergate of UFO, alien, mind–control, genetic engineering, free energy, antigravity propulsion and other secrets will make Watergate or Irangate appear to be kindergarten exercises.

Hundreds of billions of dollars already spent on the black budget can go a long way in keeping a secret a secret and in developing exotic new weapons systems while sup-pressing what's needed for the Earth. Nobody seems to want to blow the cover on the possibility that only a small number of elite individuals know much more than the rest of us know about the truth of the more tangible aspects of our tran-scendent reality—all suppressed under the questionable guise of national security. In a sense, the keepers of the secrets are right: our perception of our sense of security is threatened by the truth of a new world view. But would you want to trust the judgments and actions of a handful of individuals you do not know with the truth indefinitely?

* **The aliens**. The unfolding UFO alien abduction stories provide evidence that our visitors do not want to be found out in their quest to conduct experiments in genetic engineering and mind control. Some investigators have found evidence for collaborations with U.S. officials, so perhaps the intelligence community may have their hands tied. Perhaps we do need to give them the benefit of the doubt. On the other hand, it would seem that, since the aliens seem to be continuously warning us about our rape of the Earth, the clandestine nature of this hypothetical collaboration is in-compatible with moving toward concrete solutions. In other words, I trust neither the intelligence agents nor the aliens until I see clear signs that Gaia is truly healing as a result of their individual or collective efforts. I see no such evidence.

* **The media**. In suppressing the suppression stories (or relegating them to tabloids) the media has abrogated its responsibility to report the truth. They seem to be mostly a meek, cynical and self-aggrandizing lot, operating under strict guidelines about what to investigate and what not to investigate. Free energy, miracle cures and UFOs are intrin-sically big news, but ignored by those charged with informing

the public because of a perception of lost credibility, the safeness and party atmosphere of pack journalism, and the element of control and censorship by publishers, networks and other communicators holding the strings of our power. The current wave of UFO revelations in the Roswell story and the flurry of books and shows on the abduction phenomenon are signs things are loosening up.

* **Ourselves**. This is probably the largest single factor. At some level subconsciously, we do not want free energy, we do not want the economy to unfurl, we do not want to know we are inferior technologically or are under the control of other beings. So in abrogating our responsibility to know the truth and make the necessary changes (the most obvious of which is to restore a sustainable environment), we have unwittingly set up a power elite we do not know (or want to know) to do the dirty work of keeping the secrets. We become like children who are temporarily protected from the realities of death. In terms of money and power we pay our intelligence agents and/or aliens dearly to maintain a leaky dike, most notably our coveted and shaky economic status quo. The paradigm shift will become obvious when enough of us stand up for the truth. Then the agencies of suppression can be revealed, with forgiveness, for what they are, for the power we gave them to maintain their secrets. Then we can see these functions of those agencies disintegrate under their own weight.

Granted, the pervasiveness of misapplied free energy and alien technology transfer could lead to more doomsday weapons, so some form of secrecy perhaps should continue on the basis of "national security."

This begs the important issue of how much secrecy is justified. In my opinion, fearing weapons applications of free energy is not sufficient justification to continue with the suppression. Rather, it is a social mandate for us to mature, to go for the benefits and to disarm. We already have nuclear overkill as well as the ability to kill other human beings with guns and explosives. Disbanding the secrecy apparatus in this case would provide an example of positive intentions to use our new knowledge to benefit humankind. Also, we can probably design free energy devices to be safe to use.

Meanwhile the forces of suppression are putting such a tight lid on all this, and only we the public can decide which truths we want to know and which truths we don't want to know. If the consensus is to know our truths, we will have a less painful birth with the paradigm shift, one which I believe will result in the health, wealth and happiness for us. If the consensus is not to know the truth, we have an important political agenda to pursue.

As recently as two years ago, I was in denial about free energy. Ten years ago I denied UFOs. My bias was reflected by an erroneous assumption I made that I would have known about it if it were real and that my scientific colleagues know best. The only way I could begin to lift my own veil of denial was to see it for myself, to meet the people making the breakthroughs, to gain a scientist's credibility check of both theory and experiment. Only my own self-realization about the potential of these extraordinary happenings brought me beyond the denial. Only our collective realization about these developments can bring our culture out of its denial. As in the "hundredth monkey" fable, when enough of us have gone through that process, we will all have a Consciousness Revolution. We need a critical mass.

Thought Exercise

I invite you to reflect on a controversial new-paradigm topic about which either you or another may have been in denial. How did you (or they) avoid the truth? How were you able to go beyond the denial? Did you feel isolated or betrayed?

This brings us to the second stage of grieving—anger. Because most of us are afraid of our own anger and that of others, and because anger is considered as inappropriate behavior, we often choose not to leave our denial in the face of the obvious. Anger can also be misapplied to violence, vindictiveness, and bloody revolutions. We see signs of that happening now with the cruel bombing of a federal building in Oklahoma City, Oklahoma. We will find that during this phase of our shifts, the only middle ground left is for us to allow ourselves to feel and express our anger constructively.

4

Anger

"The truth will set you free, but first it will piss you off."
-bumper sticker by Rev. Chad O'Shea
Unity Church, Arden, NC

"When the first stage of denial cannot be maintained any longer, it is replaced by feelings of anger, rage, envy and resentment."
-Elisabeth Kubler-Ross

"The process of grief always includes some qualities of anger," Kubler-Ross writes. "Since none of us likes to admit anger at a deceased person, these emotions are often disguised or repressed and prolong the period of grief or show up in other ways. It is well to remember that it is not up to us to judge such feelings as bad or shameful but to understand their true meaning and origin as something very human."

And so it seems to be happening with the death of an old paradigm, with the Earth herself—and all the rest of us—as the terminal patients.

I am a scientist who is in the process of changing his world view. My old world view matched closely the prevailing philosophy of our time, one whose industrial applications are insidiously choking the Earth and all its lifeforms from a healthy future. To arrive at this conclusion is not easy for one who has participated in this process of destruction by insisting, with his scientific and bureaucratic colleagues, that we are enlightened human beings living under a just governmental and capitalistic system. Our basic needs would therefore be taken care of. Of course, I had been in deep denial.

I wrote in the last chapter about philosopher Michael Zimmerman's assessment of the prevailing Western world view of anthropocentric patriarchal humanism, a belief system which represents a curious blend of mankind seeking and preserving autonomy and "freedom" while declaring people (particularly men) to have a birthright that is superior to all other species. Some of this arrogance is based on the Darwinian concept of the survival of the fittest. Therefore humans are to manipulate and conquer nature by a sense of macho competitiveness.

I hope I don't have to convince you that this philosophy, implemented by all major industrialized nations of the world, has been and continues to be the most destructive force on Earth. Take a look in any city, try breathing the air, drinking the water or avoiding the sounds of machines. The Industrial Age has long since passed the point of diminishing returns. Like Dennis Weaver's example of the frog in the pond whose temperature is raised imperceptibly daily for many days until the pond has gotten too hot for the frog to move, we are gradually, insidiously filling our environment more and more with obnoxious noise, air and water pollution, solid waste, chemicals, etc. Weaver has taken the lead in establishing the Institute of Ecolonomics, building his home with recycled materials, and funding a free energy electric car.

At the governmental level the seeds of the American and Soviet military-industrial complexes sown during World War II have obviously created some of the largest blights on the environment ever experienced by human culture. In 1960, outgoing President Dwight Eisenhower had warned us about this growing bureaucratic beast and coined the phrase "military-industrial complex". Yet, in our seeming invincibility, we the people of America gave our power away to those who play on the motivation to fight aggressive foreign fascism, then communism, perhaps for good reason during the 1940s. During the 1980s, former President Ronald Reagan often suggested that the Soviet Union and United States join ranks to fight a possible new alien enemy using our Star Wars systems. Perhaps in the deteriorating 1990s we may have seen a newer enemy—ourselves.

Back in the more innocent and better-understood forties, we gave to our victorious new ruling elite carte blanche. For nearly fifty years now we have generally had the advantage of being a rich superpower with the appearance of being the "good guys," but which also has a (mostly hidden) dark side. Some of that dark side has emerged in the Vietnam War, in the abuses of executive power, sexism, and in growing domestic and militia violence. But most of it remains underground. In our collective denial, that seems to be where we want to keep it.

Our American system may have initially been the lesser of evils, but the unfolding revelations of our true nature inherent in new science discoveries would clearly render most of the Federal Government's pursuit of decadent technological initiatives such as Star Wars, nuclear overkill, NASA, DOE and Department of Defense priorities and huge industrial infrastructure obsolete and a threat to our well-being. (It could be argued that the U.S. Government is largely irrelevant to our real needs anyway, even without significant progress in the public education of new science revelations, as some of the recent secessionist movements show.)

Underneath our paternalistic bravado, we are a nation of sheep without true leadership, purpose and vision. Such vision could come from the paradigm shift but instead we see a dysfunctional denial of that vision in virtually all prominent places.

The twisted world view under which we all live is now being challenged by: (1) the appearance of technologically superior beings (aliens) who may have more control over us than our "leaders in secret" would want to admit, and (2) the prospects for the widespread use of free energy, which could conceivably render our multi–trillion dollar weapons arsenals obsolete and hence threaten U.S. military superiority. It would therefore follow that the Department of Defense and the U.S. Patent Office could confiscate free energy devices from inventors by imposing the Secrecy Act.

Zimmerman sees those in our military and secret government, motivated by anthropocentrism and autonomy, now facing their own worst fears that represent a threat to these values. Since the 1947 Roswell UFO crash, the alien presence has been an egg in the face, slipping under their helicopters and through their radars, to freely explore their installations, abduct people and mutilate animals. The result is the formation of a select few individuals sworn to secrecy with incredible powers outside the Constitution to suppress and manipulate information in a massive disinformation campaign that would make the edicts of the Roman Catholic church during Renaissance times appear tame.

If you think I'm exaggerating, I invite you to further contemplate the evidence for yourselves. The Copernican Revolution humbled us about the Earth's place in the universe. The Consciousness Revolution is humbling us about our place on the Earth and in the universe as not the only sentient beings. The Cosmic Watergates of a UFO coverup and free energy suppression represent resistances to a world view whose differences are monumental, both in scale and in significance for the future, even compared to the Copernican objections to church dogma.

The dogma this time derives principally from our support, by default, of an elite group whose agenda continues to enforce a decadent anthropocentric view that is keeping us from the greatest news in human history.

The conspiracy is cleverly cloaked in compartmentalized, need-to-know segments spanning several agencies, so most intelligence workers only know about the tiny fragments they are working on. A more coherent, bigger picture is probably known by only a handful of individuals, and they too are probably unaware about what to do about all this, given their old paradigm approaches to reality. To abrogate our own responsibility to address the truth, we have created great dangers giving this kind of power to the few.

We have not only separated church from state and made both incomplete visions of our reality. We have also separated state from state. We have created autonomous secret agencies that are accountable to precious few (and therefore temporarily fulfill the elitist goal of autonomy) and who have abused power immeasurably. By looking the other way, we Americans have abrogated our responsibility as a free and benevolent people until we can face the truth, no matter how painful it is.

It is no wonder that the Government has gone underground with black budgets exceeding $35 billion per year. It is no wonder that this apparatus has played a role larger than any of us would want to admit, this dark underbelly which has unilaterally (or in alliance with aliens), suppressed a new paradigm that clearly challenges its own power trip and its own world views—all under the guise of national security, only a part of which would be truly justified under public scrutiny. It seems that our secrecy apparatus would have only one justification for keeping things the same—if they're at the mercy of aliens.

The initial reasons for the secrecy may have been well-motivated by the need to defend ourselves from the tyrannies of a Hitler or Stalin. But in the 1990s the tyrannies of silence, of the control by the few, of escalating budgets to billions of dollars, and of a dying planet, may be far more insidious and damaging than the actions of any individual tyrant.

"We have seen the enemy and the enemy is us," were Pogo's prophetic words crafted by cartoonist Walter Kelly. We must mature in order to survive.

These tactics of denial, secrecy, suppression and debunkery have been successful, but they cannot be for much longer. For the underlying world view is now falling apart at the seams. As the recent dissolution of the Soviet Union has shown, no military power, regardless of how awesome it might become, can keep a charade going forever if its own philosophy or purpose becomes flawed. There is little integrity to this approach.

I reject anthropocentrism. I reject the manipulation of information that denies us full access to our place in the cosmos and prevents us from living our lives in balance with nature. The Cold War is over, yet the American military budget continues to soar in the hundreds of billions per year, much of it deficit spending. For what purpose? To defend freedom? Or to enforce the *status quo* by denying us our true birthright as multidimensional citizens of the universe? Or is there something I don't know about?

I am suggesting that our endorsement, by default, of this bureaucratic juggernaut, with its most egregious and insidious octopus form of control and censorship, must end or we are eventually finished as a free culture.

What is most disturbing is that, as a people, we have legitimized this bankrupt public policy by tolerating these actions. We are a culture in denial, mesmerized by the appearance of wealth, opportunity and free will. Like ostriches on a littered beach, we have allowed unprecedented secrecy and pollution to carry the day while we bury our heads in the sand as the high tide of doom approaches.

Do I feel angry? Yes, quite often. Should we all feel angry? Absolutely. In abrogating our responsibilities as a free people we have become apathetic and have *all* permitted others to abuse power we have given them for nearly fifty years.

Should this cause us to begin a revolution, consistent with these explosive new technological and spiritual changes we are undergoing now? Yes, I believe that *peaceful* revolution is necessary and unavoidable, as the tattered old paradigm falls apart. The dominoes of change are in movement. The Soviet Union's dissolution provides a living example of what will probably happen here. But I do not advocate any form of violence or vindictiveness any more than Gandhi had in India.

I have said that the mature way to handle the problem is for all of us to share in this responsibility and not to prosecute, except perhaps in extreme cases where individual civil rights have been blatantly violated. From time to time —and I believe this is one of those times—history challenges many of us to stand up where basic human ethics are involved. Revelations from the Roswell UFO incident, abductions, alternative medical breakthroughs, and free energy developments clearly fall in that category.

In the second part of this book I will be suggesting solutions on how we can make the necessary transitions to forms of activity which will be very different from what is taking place now. I foresee heart-sourced Founding Mothers and Fathers with Gaia having a seat at the round table of peace. On the placemats will be new agendas for our being, to guide us out of the tyrannies of a poisoned future we now seem to be facing. A new world–nation will be formed, based on self–evident truths and ideals analogous to those which guided our forefathers. Life could truly become a feast of love, a pie of abundance equally shared.

We will probably find a need to rebuild our house entirely, while reinforcing the basic values reflected in the Constitution and the Bill of Rights. When enough of us have made our own paradigm shifts and have grieved, we can be ready for the new. I believe it will come soon principally because the Consciousness Revolution is now knocking on our door. Do you hear a heart calling to come out?

Anger and the Grieving Process

Anger is a normal phase of the grieving process and I invite you to browse through the numerous suppression stories that have spanned this curious, dark century of ours. These stories, meticulously researched by leading journalists such as Chris Bird, Jeane Manning and Jonathan Eisen, could get anyone fired up. One cannot feel but the greatest admiration for our lonely pioneers of free energy, ufology and other new science topics. These individuals have courageously carried on in an atmosphere of persecution, disinterest and scant financial resources.

As I am shifting my own paradigm, I feel the anger coursing through me as a sense of betrayal, of the undermining of cherished beliefs I once found to be true, of a patriotic sense of being part of the leading world power which once had a purpose but is now embarked on an environmentally disastrous course, devoid of a truly constructive direction. We are a nation in rapid decay, a sinking battleship without a rudder, obsessed with the suppression of informa–

tion that inevitably leads to a world view which we will soon need to embrace like a forgiven child, in order for the whole family to survive in the long run.

And so I say from the perspective of anger: let's scrap the old philosophy and let's scrap the old bureaucratic beast. Let's eliminate unnecessary secrecy! Under natural law, the United States Government is not performing well now. We will need to let the people decide that and take charge, just as Soviet citizens did suddenly and unexpectedly just a few years ago.

But let's not throw out the baby with the bathwater. We need to look at what is good about our Constitution and Bill of Rights. The founding fathers formed our government to be free of the tyranny of the few. It was to be a government by the people, of the people and for the people, with all civil rights guaranteed.

But the Government has slowly eroded to a point of decadence which reminds one of my Italian colleagues of the Roman Empire, while it keeps from its own people information that is of great value to their wellbeing. The Clinton Administration's well-intended efforts at reform were but a drop in the bucket of what future government will need to be. Since it seems special interests resist even the most modest of reforms, they will need to surrender much of their power as well as the existing Government itself.

I come to these conclusions reluctantly after years of research and exposure to many of the principal participants on both sides of the paradigm fence. I do not take my new responsibility as an advocate for a peaceful scientific and social revolution lightly. I sometimes feel like a Paul Revere, a heralder of a new kind of patriotism. (Societal anger is now being expressed in the recent, deplorable incidents of Waco, Texas and then Oklahoma City—ironically both on Paul Revere Day.)

My anger first appeared a few years ago when I witnessed the death of purpose of NASA. Originally chartered as a civilian agency dedicated to the open exploration of the universe, administrations have recently appointed military officials to top positions in NASA and the agency has conducted several missions in secret. I have witnessed its illegal, unethical practices to cover up and officially ridicule evidence about the Mars anomalies, for example, as shown in *The McDaniel Report*. The spectacular failures of the Challenger, the Hubble Space Telescope, Mars Observer and Space Station plans are but the tips of the icebergs of underlying fear and greed.

And I have seen the Department of Energy decay into blatant disregard for our environmental mandate as if it were a wistful memory of the 1970s, as we engineer our fossil fuel and nuclear devices toward our own oblivion. In short, I have lost patience with most all of our governmental institutions. They must change, we must change.

We are not a free people. We are paying our taxes and directing our energies and attentions to outworn priorities. We are held back from our true potential by not opening up to the new information. Most of us have jobs we don't like and have little or nothing to do with our wellbeing. In principle we can live in harmony with nature, but we can only do that by rejecting the current system, both public and private, and replacing it with another system. We can retain our individual rights of free expression while doing service for the greater good, based on self-evident truths, just as our forefathers had declared.

I believe we shall find that, for minimal labor and environmental costs, we can take care of our most basic needs such as food, clothing, energy, transportation and shelter without any needs for a cancerous infrastructure that is eating away at our very essences. I will be making some specific suggestions about this toward the end of this book.

Whatever happened to the warnings of Rachel Carson in *Silent Spring*, Paul Erlich in *The Population Bomb*, Charles Reich in *The Greening of America* and other prophetic perceptions of the counterculture of the 1960s? I feel sad and angry that little or nothing seems to have been done to heed those warnings. But at least two things are different about our 1990s: (1) our environmental problems are far greater and awareness seemingly less, and (2) some of us are finding solutions that demand an entirely new perspective.

I consider my anger to be healthy. It has moved my energy beyond a denial based on intellectual safeness, smugness and complacence.

In rereading my own material of recent years, I notice a continuing respect for the powers–that–be and the old paradigm. I had a wistful hope that the new could be incorporated into at least some basic aspects of the old. I found my ongoing work to still contain elements of denial in which I implicitly aligned myself with anthropocentrism while cautiously pushing the edge of the envelope toward the transcendent wonders of the universe.

Alas, I was still using the old world view to describe elements of the new one and found myself curiously adrift, while experiencing occasional feelings of anger, yet still bargaining with the old. I found I could not yet come to grips with the data which didn't fit contemporary world views.

Of course, all this is grieving, a process which for me has been going on for several years. I am finding as I let go of these successive layers of grief, I am becoming freer of the old. I only hope your process will be easier and more rapid.

Thought Exercise

Again I invite your participation. Reflect on your initial feelings when you moved through denial and then anger in your example of a paradigm shift. Did you or do you still feel angry or betrayed about not receiving the whole truth?

Or do you remember times when you tried to restore a former peace to your situation by being on your best behavior? If so, you have probably also experienced the bargaining stage of grief.

"The third stage, the stage of bargaining is less well known but equally helpful to the patient, though only for brief periods of time. If we have been unable to face the sad facts in the first period and have been angry at people and God in the second phase, maybe we can succeed in entering into some sort of agreement which may postpone the inevitable happening: 'If God has decided to take us from this earth and he did not respond to my angry pleas, he may be more favorable if I ask nicely'...The bargaining is really an attempt to postpone; it has to include a prize offered 'for good behavior'...Most bargains are made with God and are usually kept a secret."

–Elisabeth Kubler–Ross

5

Bargaining

"The appropriateness of your diligence (snort) should be exceeded only by the prudence of your devotion (snort) to your tasks in fiscal matters (snort, snort). It is therefore incumbent on me to remind you (snort) that I expect success to be achieved by your diligent support (snort) of those infrastructure elements which ensure the appropriate expansion (snort) of anthropocentric, paternalistic, pusillanimous, perspicacious, pugnacious, parsimonious pertinence (snort) that presupposes piglike, petulant, prudent pecuniary practices (snort), as would be befitting of optimal profits from the perspicacious tendering of monies (snort, snort)."

–The Lapis Pig, 1994

None of us enjoys being angry all the time. Elisabeth Kubler–Ross' research on grieving imminent death shows that the anger that lifts us out of denial often gives way to bargaining. In this phase of the process, one wants to try to reconcile the loss (or any startling new finding like free energy) with more familiar frameworks of understanding, to seed a safer, older way of being. Culturally we would call that familiar framework the old paradigm.

During the four negative stages of grieving (denial, anger, bargaining and depression) we can begin to recognize that most humans show these four distinct behaviors more or less in sequence. Whereas anger and depression represent feelings, denial and bargaining are defense mechanisms designed to cope with the loss during those times when mere feelings don't seem to do the job.

The reverse can happen too. Anger can take us out of denial. Analogously, depression is a common feeling which takes over when attempts to bargain with the old do not succeed in bringing in the new. While none of these steps could be considered as rational or pleasant, they appear to be necessary to complete a grieving cycle. Therefore I have found it wise to respect these feelings and behaviors.

In the end, bargaining does not work. It is a temporary retreat formed on the cusps of anger and depression, a holding pattern akin to denial that keeps us away from our feelings. It doesn't help in directly bringing in the new any more than it can help bring us back to life. Yet it can still be a useful exercise in paradigm–testing. I personally confronted my own bargaining behavior upon arriving at the Stanley Hotel in the front range of the Colorado Rockies last spring for the second annual Institute for New Energy think tank/retreat and public symposium.

Estes Park, Colorado, May 9–12, 1994

Meredith and I arrived at the Stanley Hotel in a red sunset, time–shifted from our hotel departure in Italy just twenty–four hours before. The Rockies looked pure and majestic. The old hotel is a 1920s white frame building with a cupola made famous in Steven Spielberg's film *The Shining*. The hotel may have appeared as a welcome haven, but our relief was only short–lived.

The front desk managers informed us they didn't have our usual selection of a room with a view because they all had been reserved for the crew of Jim Carrey's latest film *Dumb and Dumber* which would be shot throughout the hotel. We were told that all this would be going on at the same time as the coming three–day think tank among free energy inventors, scientists and engineers. Moreover, our baggage had not connected in Frankfurt, with no promise of delivery for at least another day.

That next day, when the eerie light of a mostly eclipsed sun shone through the windows of the meeting room, I discovered I was the only attendee out of fifty who had been passed over—no registration packet, no place in the discussion agenda, and no place on the program that had been promised for the upcoming New Energy Symposium to be held in Denver just after the mountain retreat.

In what was going to be a conference of major historical importance in the development of free energy, I suddenly found myself in a nightmare of feeling like a nonperson, of losing my identity as a scientist and organizational founder. My return to the North American continent was laced with feelings of disappointment.

As the meetings progressed, many of these imbalances began to rectify themselves. Still the blow to my (then) fragile ego and my travel fatigue made me want to run away, to do right–brained things like play the piano, socialize and hike rather than be ensconced in a series of high–powered technical and political discussions with colleagues. Meredith and I thought how ironic it was that some of the smartest people in the world were gathered here alongside Hollywood characters in the filming of the frivolous fling of *Dumb and Dumber*. This synchronicity of opposites set a stage of welcome comic relief to a serious happening.

Under most normal circumstances the meeting would have engaged me. Sitting at round tables during the meals would have been an opportunity for me to catch up with the superstars of free energy such as Thomas Bearden, Moray King, Paramahamsa Tewari, Shiuji Inomata, John Hutchison, George Hathaway, Toby Grotz, Pat Bailey, Don Watson, Win

Lambertson, Harold Aspden, Hal Fox, Stefan Marinov, Tim Binder, Dale Pond, Ken McNeil and Don Kelly.

During the mornings, afternoons and evenings, the fifty of us, virtually all men, would go into session sitting around a large horseshoe formation of tables while we discussed research results and strategies for the fledgling Institute for New Energy (INE). Several years ago, Maury Albertson, Colorado State University professor of civil engineering, optometrist Robert Siblerud and I had founded the International Association for New Science (IANS), parent organization to the INE. Our discussions centered around birthing the INE in a way consistent with the purposes, charters and personalities of the two organizations' leaders. Disagreements between the two groups about the conference organization resulted in passionate feelings, of which I was to become a mediator and then a provocateur, as in a Shakespeare play. These political events were later to unfold during the conference. But of most interest during the first day of the think tank was the revelation by leading industrialists of what kind of marketing plan would await us.

Bargaining with Industrialists

Our conference financier presided over much of the meeting. This one man is now worth about $200 million from selling his software company. True to his calling as a suc- cessful entrepreneur of technology, his interest in free energy goes beyond his spending tens of thousands of dollars to fly in the best and brightest researchers with their hardware. This wealthy man has the potential of becoming the J.P. Morgan of free energy.

It was at this point during the think tank when I began to become interested in what was going on around us in the hotel. As the crew of *Dumb and Dumber* was blowing artificial snow from machines outside the hotel, cranes with cameras were pointing down through the large arched windows of the conference room, perhaps as if the CIA became like Keystone Cops infiltrating the meeting. Looking out another window, we could see the front porch pillars were decorated with Christmas garlands, and expensive cars pulling up and bellhops walking through the fake snow on this balmy May afternoon. Nevertheless, I listened intently to the conference sponsor as his marketing people began to explain the rationale for their vision on how industrial free energy might unfold.

The interest level was high for all of us for obvious reasons. Most of us in the room were well-seasoned at struggling with paltry funding, ridicule, suppression and the lack of the kind of team effort that will eventually be needed to develop free energy. And yet the financier's marketing vision did not seem to meet any of these immediate needs. In other words, the profit motive was not necessarily aligned with the motive to develop free energy quickly.

He explained to us that the situation was similar to that of a river. For optimal profits we want to catch the river where and when it flows fastest. Until then it would be foolish to put up much money now. Why spend millions on researchers now, with uncertain results, if we could hold back those millions until such time as a device is truly ready to enter the marketplace and can outcompete the others?

And so, in throwing only small bones toward free energy research and not going for a commercial push at this stage, this industrialist appeared to be joining the ranks of the Government and existing corporate infrastructure, waiting for the other shoe to drop. Meanwhile the inventors continue to be neglected, underfunded, persecuted, debunked and not given the opportunity to come together as an Earth Team to create what is really needed now.

Yet, in a sense, I felt intrigued by the marketing scenario. After all, it did answer the concern that an existing powerful system, industry, could be up to managing free energy. It certainly seemed to be a better bet than the Government. The only problem was, we would have to wait until the winning horses could be identified and lined up at the starting gate. Then we could begin to manufacture them. All this would optimize profits for the owners and not misdirect resources to potential losers.

But all this is old paradigm thinking. Can I bargain with that? Can the Earth bargain for that?

I found myself caught in a dilemma. On the one hand, the symposium sponsor seems to be an honest and successful man who took on an important initiative. He has funded these very important early meetings which probably would not have been possible without him. On the other hand, he represents a point of view that places profit, control and the competitive edge first.

In confronting this dilemma, I began to learn at a deeper level the two sides of the issue. Can industry, backed by its own claim of greater efficiency over Government (the only apparent other show in town), be trusted to develop free energy without price-gouging and insensitive environmental planning? Or why not just let the marketeers do their thing, to strategize at their gaming boards waiting for the river of optimized profits to run swiftly, poised for the winning moment to come? The world could certainly wait awhile longer. Or can it? And if the American businessmen don't do it, the Japanese surely will anyway, right?

These conflicting feelings left me both in awe and anger with the sponsor. Yet, wasn't the whole exercise for me but one of bargaining? I began to find I could not continue to exist with integrity within given industrial systems—both the current ones and those being hatched for the future. Yet this money man had done a great service in bringing people together. What was his real agenda? There would be no better way of finding out than to confront him.

And this is exactly what I did, although somewhat inadvertently. "Should we consider the possibility," I asked him, "that a government or some 'angel' fund a free energy R&D laboratory to the tune of about $100 million so that we can get this thing on line as soon as possible?"

I feared, by his coldness toward me, he did not appreciate my question. I presented to him what seemed to be the most sensible option, but one which might steal from him his potential control of the new technology. While my intention was to stimulate discussion of alternative strategies for implementing free energy, the result unexpectedly and (in retrospect) understandably fell flat on the ears of those who had different agendas.

But the free energy waiting game was not only reserved for governments, industry and academe. I have since asked leading financial managers how they felt about the timing of the introduction of free energy. Probably not for at least twenty years, according to a cousin who is president of one of the largest banks in Albuquerque. After a speech I delivered there on the prospects and implications of free energy, my cousin came up to me expressing intrigue about its potential. But it was also clear to him that the current banking and financial systems could in no way handle the vast changes precipitated by free energy for at least two decades because of the need to reconfigure such major sectors of the economy as the oil companies and utilities. He also acknowledged to me that the economy is a delicate house of cards that could fall anytime.

Again, with mixed feelings of awe and anger, this time toward my cousin, I pondered, bargained and lamented. Does this mean we have to suppress everything for another twenty-plus years while the powers-that-be figure out ways to shuffle their own self-interests, under the "lapis pig" pretense of protecting stability in the economy and jobs? What if the Japanese manufacture billions of free energy gizmos for a dollar apiece anyway? When will we learn? Must I be nice and bargain with this kind of consciousness in order to be "realistic"?

The Politics of Bargaining

Getting back to our conference in Estes Park, I also became caught in a political crossfire between factions within the governing bodies of the parent and child organizations sponsoring the events. But on the next day, when I perceived that the younger faction was prematurely moving ahead with its own power structure and election proceedings, I stood up and defended the point of view of the parent organization, thus delaying proceedings for two days. This was an unfamiliar role for me—that of a critical parent.

As we all left Estes Park and headed down the mountain to our public symposium in Denver, I was one of the least popular among colleagues. At a deeper level, I began to learn that bargaining does not work. Compromise, yes; bargaining, no. The truth of the matter is, the free energy research community is in dire need of support and of an environment of togetherness. It must be a team effort, yet allow individuals to develop their own creative talents. We have been hiding things for too long.

The process of coming together has barely begun. One way or another, we are going to have to find a way to make this happen, and I would be willing to compromise on that—whether it be industry, government, a hybrid, or some totally new entity. But I wouldn't compromise or bargain with anybody that puts their own self-interest above the greater good—hence the need for altruism at the top.

I began to reflect on what bargaining had done for me earlier in my life. I had often gone from one old paradigm environment to another hoping that each new one was the real one, only to find out later it wasn't. And so it goes for all old paradigm systems—Government, industry, money managing and academe alike. How could I bargain with systems that I don't seem to fit into or that put their own profit well ahead of humanity's? How can we evolve new systems to better accommodate the Earth? How can the Institute for New Energy help? Posing these questions eventually led to some new ideas which will be presented in Part II.

I also reflected that human politics is normal among a growing new community which could become influential in the near future. My choice to enter the web was mine and so were the consequences. Others were involved too. As pioneers of the unknown, we are all challenged to wonder what our new identities will be, having cast off into the void. Therefore, many of us might have "feisty" personalities unwilling to bargain with any old ways of doing things.

Bargaining is where we desperately want to be on our best behavior and to find ways to fit into the old. In full action this approach seems to be oddly inappropriate. There is a vague, inconsistent, humanly frail quality to it all. A bargainer is one whose paradigm is shifting and is therefore having trouble fitting in. He or she will be likely to be found out soon, discounted by the insiders and destined for depression. And yet all this seems to be part of a healthy grieving process!

From Bargaining to Depression

I have mentioned that for eighteen years, Bruce and John Klingbiel, Christian Science practitioners and outstanding researchers on the power of prayer, tried to convince their own church of the value of their work. Instead, they were dropped from the church practitioner lists (their livelihoods) and basically excommunicated. Their work was also not accepted by parapsychologists because of difficulties in communicating their concepts. I described in *The Second Coming of Science* this dramatic case of a father–and–son team who clearly entered the new paradigm, but still tried to bargain with the old by seeking the approval of both the scientific and spiritual communities.

The results were disastrous. In not having their work accepted by their peers, the Klingbiels slid into a deep depression. About two weeks after I met them for the first time, in April 1993, they went into the woods and were later found dead, shot in their heads, an apparent double suicide. They had bargained for too long with the old systems; their own mental health had declined and they had become modern–day martyrs. Would I do that too? Not if I could help it.

The cases of the Klingbiels, of free energy inventors, of UFO researchers, and of new scientists in general, underlines some of the pitfalls of grieving if we do not know what we are doing. Most of us are still in denial and hanging on to the old paradigm. When some of us move into anger, that is usually a good sign we are starting to embrace the new, but anger is healthy only as long as it is constructive and not hurtful to the self or to others.

Bargaining can also be a positive behavior if it is short–lived. It is a method of exhausting remedies, of eliminating old paradigm structures to advance new paradigm concepts. If we are lucky enough to discover early on that bargaining does not work and that you, the rebel, will not be accepted by the old, then the grieving can go on. Reflecting on my own career, I had been bargaining for two decades before I learned I simply didn't fit full–time into any existing organizational structures. And I live to tell the story!

The key to all this is examining the degree of alignment of purpose between ourselves and our potential cohorts. Bargaining is more a mental process than a feeling. If, for example, the dominant agenda for the INE or any other organization were to support existing industrialists in commercializing free energy, I am out. I wish them Godspeed. If, on the other hand, its principal purpose is to bring together like-minded inventors, scientists, engineers and informed lay people for the purpose of developing free energy as soon as possible to restore sustainability to the Earth, I am in. Which way any organization will go is up to us, the members. The jury is still out for the INE.

I have long since abandoned bargaining with the policymakers of our existing governmental departments regarding free energy, UFO and new science issues. They have consistently kept secrets, suppressed important developments, reached deeply within our pockets to support abusive central projects, and have already spent their own deficits like drunken sailors. They are no friends of the new science movement and probably cannot be bargained with, unless some funding could be freed up to support free energy research with no strings attached. The Los Alamos Laboratories have expressed interest in free energy research, but I do not trust their dark side, which is deeply rooted in Hiroshima-type weapons and other lethal nuclear technologies. Given the current Government's atrocious performance in all these matters, I give the New Energy Symposia funder and the Japanese businessmen higher grades for implementing integral free energy.

The INE is a living example of a mixed-paradigm organization, somewhat like old wine in new bottles. I decided a few weeks after the political dramas that kept me hopping in Colorado, where I had the feeling of being in a futile battle for my ego and recognition, that I needed to back off and reflect about all this. I wanted to see if perhaps my own battles could become the battles for the entire culture if we are not forewarned. I predict that once enough of us recognize the true potential of free energy and have gotten angry about its suppression, most of us will go through a period of vainly trying to fit the new within the context of the old, even though the old structures have never served the new and have actively suppressed it. Our bargaining behavior can sometimes feel like asking the fox to guard the chicken coop.

"No problem can be solved from the same con-sciousness that created it," was the Albert Einstein dictum which I have often contemplated. The actions of bargaining teach us to recognize that the new paradigm will need to come to us in totally new forms, free of special interests. This especially concerns matters of money, that (sometimes hidden) agenda of optimizing one's profits or vested interests at everybody else's expense. Such an approach is simply not consistent with the Consciousness Revolution.

The Denver part of the meeting was even more intense than the mountain retreat. Clearly it was time for me to back away from the mad scramble of male power. Not only could I not bargain with old interests, I couldn't seem to bargain with anything, with anyone. I retreated to the silence of our Oregon home feeling deeply despondent about my own lack of identity and at the powerlessness of our species to come to grips with its greatest challenges. I realized I was entering the crux of my final negative stage of grieving: depression.

Thought Exercise

And again, I invite you to reflect on your own example of a paradigm shift. When bargaining and anger didn't work, have you ever gotten depressed about it? If so, where has it gotten you?

"When the terminally ill patient can no longer deny his illness, when he is forced to undergo more surgery or hospitalization, when he begins to have more symptoms or becomes weaker and thinner, he cannot smile it off anymore. His numbness or stoicism, his anger and rage will soon be replaced with a sense of great loss."

–Elisabeth Kubler–Ross

6

Depression

"You're in the machinery room, and when you're in the machinery room you will be spanked and sent to the electric jail!"

–Brian O'Leary, about age 6, uttered repeatedly

Ashland, Oregon, May 21, 1994

Blue patches of sky across the valley show signs of clearing, and my depression continues. It is the third day home now in our retreat in the wooded mountains of Oregon, and a dreary drizzle and temperatures in the forties have been with us ever since our return, even though it is late May. But the air is clean!

I am back from another disorienting six–week whirlwind of conferences, speeches, and visiting various scattered free energy and UFO scientists. In the past four months I have been speaking or meeting with people in Bombay, Bangalore, Tokyo, Rome, Florence, San Marino, Los Angeles, San Diego, San Francisco, Lincoln (Nebraska), Honolulu, Kona, Kauai, Seattle, Estes Park, Denver and elsewhere. Almost everywhere I have breathed and choked in the foul air of a dying planet. I see no recognition of future possibilities of doing anything about it from either the mass culture or our leadership.

Just two months ago, in our peaceful woods, the absentee owner of the adjacent property authorized his land to be logged. The rape of the beautiful forest of fir, pine and oak caused Meredith and me to cry over the loss of an enchanted glade we used to visit. One day I walked over to the near–clearcut wasteland of stumps and fallen branches, now being prepared as a homesite, put my head onto the ground, and cried. I began to empathize with the Native American position. I started to pray to the Earth, please help me help you. Or, when the depression turned to anger, I would declare, HELP ME HELP YOU!

Nobody likes to admit they're depressed and nobody wants to be around a depressed person. So I usually cover up depression with distractions and food and alcohol and our daily rituals, stuffing the depression and thinking perhaps it might all go away at some indeterminate future date which never seems to come. But my bouts have deepened and now there seems to be no escape from the depression.

This time I have decided to ride with it and use the writing of this book as a tool that might move the negativity through me. Then, perhaps some inspiration will come. Maybe my energy will be freed, all our energies can be freed, and we can begin to enter a Golden Age.

I have every reason to feel depressed. As measured by income, I have had two of the most unlucrative years of my life, barely making monthly payments and stretching credit cards to the max. My speaking and writing career in new science has been on hold while I scramble doing twice the work for less than half the income. I feel I may have lost the respect of some of my family, friends and colleagues. Why so? Have I moved too far away from our mainstream culture?

I think about the Klingbiels and their probable suicides. Would persecution and death also be my fate, a modern-era martyr such as a Galileo or Bruno? Or could I somehow shake loose these shackles and move ahead into the new paradigm?

I feel truly at the doorstep of despair, of wanting to give up on living. I see myself chronically depressed. Occasionally, when the pain is too much, I revert to depression's flip side, anxiety and anger, to avoid facing the despair. I am emptying out, feeling like a nobody and an unpleasant chap to be around. But, thanks to this awareness and to Meredith, I am really able to get into these feelings without much recrimination at home. I see how we may all need to descend to this level of grief, a paralyzing depth that seems to embrace all time and space.

And if the travels weren't enough to cause us to stay still, we have returned to managing a jumble of ambivalent future marginal events and checks written to us that have bounced. Keeping up on house payments is a challenge. All this is no surprise. Chaos seems to have taken over my business as well as my nightly dreams. Even this book has an uncertain fate as a result of my backing out of a New York publisher's demands that I stick to the "facts" and write a book entirely in the third person.

In spite of all this, Meredith and I still want to express our vision of a better world, for us and for everyone. But the dreams seem to wither more and more as the treadmill patterns of our daily survival begin to dominate. Memories of happier times, visions of global transformation, seem more distant than ever.

Yet paradoxically, psychologists tell us that sometimes the closer we get to our goal the further off it seems to be. When can we ever emerge from this nightmare, I wonder? Or more recently, the question has been, *will* we ever emerge from this nightmare?

Lacking a coherent business relationship with the outside world, the only thing we seem to have remaining is each other, our deepening relationships with certain others, and the knowing that we are going through a paradigm shift—personally and collectively. We would have to wing it, the future is up for grabs. No more retirement plans, no more medical insurance, living on the edge...of what?

My main sign of hope now is that depression can be the final dark before the dawn, a necessary step in the grieving process for old outmoded ways. The universe seems to be telling me that denial, anger and bargaining would no longer work. I am now in depression, and I guess I had better feel it in order to get beyond it. Maybe this depression would be the last one, maybe the process of writing this book could help move healing through me once and for all. That is my wish. Otherwise, how could I ever help the Earth if I just keep on feeling depressed?

Home, Ashland, August 2, 1994

The sky is smoky with forest fires that could come here anytime in this drought of droughts. I am now hearing the quintessential sound of the electric jail, the sporadic blare of an electronic tone issued by a neighbor from a horn pitched about an octave below middle-C. What a huge distraction this is from the whispers of the trees rustling over our deck.

The noise has been with us for a few days now and has moved us inside the house until we get a response from our protest. What was that device making all the racket? The new scientist in me was curious, but the more basic feeling was anger toward a basic invasion of my space—noise pollution. I later discovered my neighbor used the device to fend off deer from his vegetable garden. As a musician, I am supersensitive about noises that we are creating like fax machine beeps that can't be turned off and trucks that back up with the obligatory revolting beep–beeps, and the endless roars and humming of internal combustion engine motors, power mowers, power saws, leaf–blowers, etc. Whatever happened to ice cream truck jingles or San Francisco fog horns? The frog in the pond comes to mind.

As I look back at the words I wrote last May 21, I begin to see the patterns of the human quality of depression. My own depression is a classic case. So what should I do—go out and buy some Prozac? No! The depression and the void must be entered. If I don't do that, I'm missing an important step and probably will need to go back to denial and bargaining (buying out again for temporary gain) and/or anger (you son of a bitch!).

The good news is, depression is usually the last of the negative stages of grieving. I am beginning to learn that the initiation of depression is the culminating rite of passage in the drama of life. *Depression harbors that moment of truth when we face our worst demons before the dawn. Death, and then Light, await at the other end of the tunnel. We have no choice but to move beyond, but first through the void.*

I am also deeply disillusioned with just about everything connected with my identity and the ways of the world; disillusioned by scientists, my former colleagues, who have abandoned the quest for truth; disillusioned with the deepest levels of arrogant human behavior not even befitting of a lower animal; and disillusioned with a possible failed human experiment run amok while our alien genetic engineers come back to try to fix things under conditions of stealth. Perhaps all they think of our species is that we're a bunch of laboratory rats who have gotten out of their cage, and some of us need to be put back in to be reengineered.

I have recently been handed a dossier by a trusted colleague of a multiple personality case of a woman who claims to have been a victim, as a child, of torturous mind control experiments by various secret Government agents along with aliens in an underground base outside of Las Vegas during the 1960s and 1970s. Shortly after giving a public presentation of their story, she and her husband disappeared! What happened?

The story seems incredible at first blush, but hangs together with other eyewitness accounts. Perhaps we have another even more bizarre Roswell in the making. The alleged child abuse is shocking enough to have sent both my colleague and myself into even deeper depression. My colleague had been a career Air Force man, a Ph.D. hypnotherapist who had helped prepare our men for battle in World War II, Korea and Vietnam. I had been an astronaut. At a recent meeting convened by Laurence Rockefeller, I met several other former and current military and intelligence officers who were witnesses to UFO phenomena. They feel the same. What does this kind of information do to our sense of patriotism?

I sometimes feel like living like a dropout of the nineties, thirty years overdue from a lifetime of career-building. For what purpose did I serve? I seem to approach this book and Meredith her painting *The Last Supper of Gaia*, as if they were our last projects. Apocalypse seems to look like an Armageddon lurking just around the corner, waiting in the dark shadows to consume us.

Types of Depression

At the time I had written these diary entries I had been unaware that Elisabeth Kubler–Ross identified two types of depression for those in grief—reactionary and preparatory.

The reactive depression naturally follows from the empty feeling of loss after nothing else seems to work. According to her research, individuals suffering from this kind of depression can often move through it quickly, if we can help them cope with shifting their focus from their traditional (old paradigm) responsibilities and activities to their newer situation or impending death.

"We are always impressed," she wrote, "by how quickly a patient's depression is lifted when these vital issues are taken care of." Societally, this means helping to meet the survival issues of those who have been displaced by the enormous inevitable changes that lie ahead.

Preparatory depression, on the other hand, is "a tool to prepare for the impending loss of all the love objects, in order to facilitate the state of acceptance." She goes on to say, "The patient is in the process of losing everything and everybody he loves. If he is allowed to express his sorrow he will find a final acceptance much easier, and he will be grateful to those who can sit with him during this stage of depression without constantly telling him not to be sad. This second type of depression is usually a silent one in contrast to the first type..."

Looking back on my own diaries of May and August 1994, I seem to have been afflicted by both types of depression. At first it was clearly reactionary, which would make me elicit a statement such as, "I have every reason to feel depressed." Such a depression can be moved through with help from others, as well as from using it as a springboard to redefine our roles and responsibilities here on Earth.

Later, I began to feel some aspects of the second type of depression—a quiet and lonely entering of the void of death that can lead to a final acceptance of our situation.

Again, I invite you to reflect on any feelings of depression you may have had around your subject relating to the paradigm shift. Do you feel you may have moved to a stage of a more silent, preparatory depression, which could lead to an acceptance of our impending, collective death of an old order?

Home, Ashland, February 1995

I often reflect on the human condition. How can we as a race imbue ourselves with such complimentary qualities as intelligence and sentience, when from a more cosmic perspective, we have built an anthill which has taken over our Earth? How could this race have organized itself to bring us to the point where, like the Borg in Star Trek or a Dyson Sphere, we are on the brink of consuming and condemning our habitat and its inhabitants to death?

The problems we now have have run the full gamut—not only health care, pollution, military and energy abuse. We have festering pockets of famine, malnutrition, unchecked disease, overlogging, crime, wars, domestic violence, lack of leadership, and a continuing sense of anal-retentiveness from the elite. Our highly coveted economy and technological base have become our gods, our sacred cows that precede and preempt all other approaches. Our dominant hidden agenda is the acquisition of money.

"Science and technology," writes futurist Willis Harman, "have raised human living standards for millions beyond all expectations, but social inequities, political stresses, and unreflective uses of technology have polarized humanity and degraded nature. They are creating problems of planetary dimensions. Global warming; the attenuation of the ozone shield; the menace of deforestation and desertification; the destruction of many species of flora and fauna; the extensive pollution of air, water, and soil; and the poisoning of the food chain are threats that all of us share."

I reflect that at the time of the founding fathers, we created a system designed to transcend political tyranny. We now need a system free of philosophical tyranny. Our contemporary tyranny is anthropocentrism. Our conceited species lacks constructive vision and instead mortgages its own future holding everybody else hostage. It actively suppresses anything that interferes with the visions we need and is destroying the planet and all other species in the process.

In his recent book *The Rebirth of Nature*, cell biologist Rupert Sheldrake has suggested that nature always finds ways to keep a diversity of species. If that diversity becomes threatened (by human encroachment, for example) it is likely that there will be sudden behavioral changes among all parties in order to effect a sort of Gaian mutation. If the species imbalance were to occur for too long, means are usually found by nature to decimate the predators by other means, such as disease. Sheldrake offers evidence for these mutations in behavior among humans and animals, and one gets a strong hint that, if we don't do something for Gaia soon, she will do it for us. It might be too late already. As we shall see later, climatologists and public health scientists are warning us that airborne diseases may soon spread fast because of global weather extremes, caused by El Nino and human-produced greenhouse gases, primarily attributable to burning oil.

Most depressing of all is the fact that it seems like few of us are able to spring out of our boxes to address the true questions of our survival and well-being. We seem to be vested in a polluting past and invested in a shaky future based on financial promises which might not be met, let alone our ecological mandate which somehow got lost in the shuffle.

I reflect on the symbol of Mars and its mysterious moons Phobos and Deimos. I have written countless articles, papers, books and theses about all that. As an astronaut, I was the only one appointed to go there, and would have been there and back by now if history had been rewritten.

Mars is the god of War. Its moons Phobos and Deimos are named after the gods of fear and terror. I was interviewed for a segment on Fox Television's show *Encounters* about the mysterious face on Mars and the associated NASA coverup. The symbol of Mars propels my anger and my depression. When will it all end?

I am looking over these diaries as small examples of the depression I now feel about my situation, our global situation. And I am also seeking a small aperture that will get me out of my depression through the void into acceptance, enlightenment, empowerment, joy and transcendence. But first I still need to address some cousins of depression: paranoia and despair.

I recall a visit to a new science pioneer who had been consistently brilliant and accurate in assessing the technical aspects of free energy and UFOs. But his behavior appeared bizarre when he began to talk about the control groups whom he believes run the planet. This strange dichotomy in personality smacked of paranoia. Whenever the subject of conspiracy comes up, I usually run for the door. After all, what does conspiracy have to do with science?

The man told me stories in which the control groups have been after him. I had also been told by some colleagues they were following me. Is any of this true? I do not know. What I do know is that paranoia is not uncommon and sometimes justified in exploring paradigm shifts. So while I do not know about the extent of conspiracy in this

Disinformation Age I do know how I feel about it: angry and depressed.

When I'm depressed I sometimes have the feeling the CIA is after me and about to assassinate me. Or I sometimes feel the converse: that I am not worth that kind of attention. In more positive moments I fantasize that those in the know will one day come forward and tell the truth in exchange for personal safety assurances and amnesty. In many ways the world and our future is hostage until we know what some people know. On the day of reckoning these agents may want to destroy evidence and run for the hills to save their skins against a vengeful public. Fine! But, for heavens sake, let's come clean.

Or is all this an illusion of disinformation webs? I know at least some of this isn't. Something is going on behind the scenes that is part of the overall suppression effort, and I am sometimes angry, sometimes depressed, and occasionally paranoid about the extent of the conspiracy of silence, which we all seem to have ordained for the sake of stability and security.

My part–time paranoia sometimes gives me the feeling I am one of the last honest and aware men in the world and I am about to be sacrificed like a Polynesian Pig. This fantasy matches well my childhood recitation of our fate here in the electric jail. To seal my fate, Meredith recently branded me on my buttocks with an iron in an accident that happened as I walked swiftly by the ironing board. Rather than becoming a normal light burn to be dismissed as a domestic accident, the wound re–opened several times as if it were active stigmata and has formed a permanent scar.

I could not understand why this was happening. Meredith, a very gentle soul indeed, swears the iron headed toward me beyond her control, like a automatic movement on a Oija Board. Making matters even more curious, here I claim to be Mr. Spoonbender–know–it–all who could walk on coals whose heat is much more intense than any silly brush with a hand iron.

I have felt myself to be in persecution for some time now. I have lost much of my personal ego, I have uncovered some shady things about myself and human power structures. I have confronted dilemmas about our future I cannot seem to resolve intellectually. I have seen the world seem to go to hell in a hand basket and have witnessed no true leadership, vision or institutional integrity. I feel remorse about being a predatory human and taking more than my fair share. I have sunk into paranoia and despair, and have emptied myself to nothing.

Despair: Facing the Void

Except for those occasional times when bullies chased me home or I was rejected by a woman, I had never sunk this deeply. Unprecedented was my frustration about my identity in the world and the prospects for the world itself. The void was enveloping me as any hope extinguished. Where do I turn?

The wise answer from inside began to speak. "Nowhere—for now," it said.

"Nowhere? Why nowhere?" I asked it.

"Because that is where you are. Be with the void. Appreciate the emptiness. For you will need to be here for awhile until you finish your grieving."

"How long will this be going on?" I asked in exasperation.

"How long do you want it to go on?" the voice asked back.

"I want this to stop now!" I pleaded.

"Then you will need to experience the void now," the voice answered. "You must find your emptiest of empty and surrender to it! Don't worry. And now I must go."

What an unfair bargain, I felt, still fearing the void I was at the edge of. Where am I? What is all this about? What does that substratum of mass-energy look and feel like? Is this void my imminent death I think about every time I get chest pains? (I had had a heart attack that nearly killed me two years ago.)

I had often heard wisdoms casually offered by metaphysical teachers, such as "you need to be willing to let go of what you don't want before you get what you want." I had once felt nothingness in a group meditation in which others described extraordinary experiences. I had left the meditation feeling a bit ashamed for my lack of "gee-whizzes."

Did my fear of entering the void mean I was still hooked on growth, devouring the planet right along with the economy, tainted with the philosophy that more is better? I began to suspect this was so. Or perhaps these exercises were designed to be a major exam, perhaps a final exam, as an Earth school initiation in which I would need to die the brave death of an ego in hopes I might some day have the strength to sprout my wings like a butterfly. But I might have to physically and psychically die first, and then what? It seemed as though my memory of the good of life had faded away.

I began to search the scientific muse for ideas about this oh-so-lonely feeling of not belonging, of being too much a part of the problem and not enough of the solution, of imminent and unresolved death.

I have mentioned that we are in a Consciousness Revolution which involves the inevitable expansion of our reality beyond our familiar dimensions. The Flatlanders had resisted the notion of a three-dimensional Earth, the

Newtonians had resisted Einstein's addition of time to the dimensional matrix, and now the mainstream resists the potential addition of new dimensions which we can just begin to visualize and understand.

Earlier this century, the great physicists Albert Einstein and Sir Arthur Eddington proposed that the universe is shaped like a donut with an infinitesimally small hole in the center. The formula of the surface is also that of a four-dimensional sphere called a hypersphere, represented in three dimensions by the donut with a tiny hole in its core.

This archetypal form was noted by the late inventor-philosopher Arthur Young to explain a variety of other phenomena including the vortex and perhaps consciousness itself. A hypothetical circular fence on top of the donut appears to be insurmountable, yet if one of us could find the tiny hole and pass through it, he can come over to the other side. Finding that hole and having the courage to pass through it might be akin to using the inner consciousness to blend and unify with all consciousness.

Young went on to suggest four separate dimensional structures underlying the hypersphere or donut, which include seven kingdoms of existence. The lowest dimensional structure contains the minerals, in which energy and position are both known and determinate. Objects in this state have no degree of freedom and maximum limitation.

The next structure embraces the plants and atoms, both of which have one degree of freedom: energy. The third structure comprises the animals and nuclear particles, who have the additional freedom to move around in position. The fourth includes humans and perhaps the higher realms of light, who have the greatest flexibility to transcend limitation.

This sequence represents an evolution from the kingdom of greatest constraint and density (minerals) to those of maximum freedom (humans) who are evolved enough to potentially reach into spirit and be unified with the whole. The key appears to be for us to be willing to find the hole and to risk the dive into the mystery of its black nothingness. Perhaps this "death" could bring new light.

Overviewing the Paradigm Shift and Grieving Process

We are a culture in denial of an unfolding reality. Sooner or later, we will need to get beyond that denial if we are to survive. The old world view of anthropocentrism, outdated Cold War politics and optimizing profits have resulted in the abuse of industrialization, the rape of the Earth, and a powerful predestination toward continuing these practices indefinitely.

The new realities of free energy, alien contact and alternative medicine threaten the old world view, and so are suppressed in all major areas of our expression. Because of growing activity and interest in addressing these topics I believe the suppression cannot last much longer. History teaches us that radical changes in science and technology provide accurate early warning systems for the transformation of the culture as a whole.

I believe we are in the midst of a Consciousness Revolution that few of us are now aware of. The essence of this enormous paradigm shift is based on our empowerment to influence the material world, our potential to use energy cleanly and appropriately, our willingness to embrace the alien question, and our ability to heal ourselves and others free of medicine and surgery. These developments will profoundly affect each and every one of us in the near future, because we have no other choice. "Shift happens" is one of my favorite bumper stickers nowadays.

When the systems we have created take on greater importance than the plain truth, this is the stuff of big change.

New science developments ensure us that extraordinary technologies await us on the other side of the paradigm gulf. Prophets have also been telling us about the coming millennial changes. May we quickly become wise enough to enter our own grieving cycles, so we may become enlightened and empowered to take on our new lives with joy!

Because the changes are operating on such a deep level we will all have to grieve the old as if we were facing imminent death or losing a loved one. Clinical research indicates we cannot be healthy or come into balance until we have gone through the stages of grieving, which includes denial, anger, bargaining, depression, and acceptance.

In Part I, I have modeled the first four stages of the grieving of the old paradigm, which I have personally experienced over the past several years. Because I had ascended to positions of power within the context of the old, I have felt a lot of pain during my process of leaving and grieving the old structures. Elisabeth Kubler–Ross writes, "The rich and successful, the controlling VIP, is perhaps the poorest under (grieving) circumstances, as he is to lose the very things that made life so comfortable for him."

Again, I invite you to examine your own feelings of grief around the coming changes. You may even be able to track where you are now in the process. For example, I find myself at this moment on the cusp of depression and acceptance, with occasional revisits to denial, anger and bargaining along with new visits to enlightenment, empowerment, joy and transcendence. I am wanting to complete the negativity and to go on to the positive aspects of change which await me on the right side of the table of Gaia, to the stages of celebration. But I have had to empty myself of much of my ego and material attachments. I have had to feel a lot.

In our patriarchal culture, we have placed high value on being cool, rational and nonfeeling. These behaviors are just the opposite of what we will need to do to go through our grieving. This poses a problem, because the ones who will need to do it most intensively are the ones least likely to want to do it because such behavior is considered as inappropriate. And so most of the men in power are caught in the trap of denial, hanging on for dear life, not wanting to acknowledge the new, and not wanting to express their feelings, to do Robert Bly's ashes (grief) work.

The problem is, all this could be a ticking time bomb. The history of revolutions is replete with examples of vindictiveness and overzealousness from the outside and backlash from the inside. These behaviors represent a destructive anger that could destroy us as we see violence in America spread. The paradox is, that which we need to express most is the very thing that could bring us down, if the anger were to become public policy. Our fears of the anger that await us beyond denial are keeping us squarely in the denial, even as the emperor is removing his clothes.

How can we get out of this dilemma? The answer for me is that we first acknowledge the grieving process we will all need to go through. *We need to support and respect all those going through it—even those in power whom we might be angry with. We cannot let anger dictate our actions toward others but we must be able to express it.*

The same idea goes for bargaining and depression. If we can face and accept the fact that it is healthy and necessary for all of us, regardless of power and position, to feel our feelings of grief, I believe we can all get through these seemingly irrational times. Everybody needs to be supported through the changes.

From my own experience, I would suggest making the acceptance of the truth, as well as acceptance of the rather bizarre state of our culture, as our early goal. It is a more balanced state, similar to being at home with whatever presents itself, and a knowing of a transcendent truth. If we cheer one another along the process of grieving, more and more of us will get into acceptance, and soon the world will change. It might be that simple!

Part I Epilogue

"For the meek will inherit the Earth."
–Jesus

Puttaparti, India, February 15, 1994

We return to the story at the beginning, the private interview I had with Sai Baba. You recall my enlightenment about the importance of consciousness that precedes the manifestation of matter and energy. This realization followed some dramatic examples of both manifestations that had just occurred on exotic Indian soil. We now pick up on the conversation we had on this extraordinary day.

"How are you?" asked the clairvoyant swami. "How can I help you? I see you have been depressed, yes, very depressed much of the time. So much worry about money, about work, about your health... Don't worry, be happy! You have great future, good future! You're a good man, good man!" His hands clasped mine as we looked into each other's eyes.

"Really?" I thought about the gold watch Sai Baba had produced for me the previous year, which I have seen as a retirement present—retirement from worry.

And so with those thoughts and feelings in my heart, filling the empty void, Kubler–Ross' preparatory depression, I began to accept the higher wisdom of my own situation and the state of the world. I began to surrender rather than to resist, and I began my free fall into the hole of the donut, the void that precedes the New World.

Meredith Miller 95

Prologue to Part II

Free Energy Magnetic Magic

The new currency
is shared energy
as each point of consciousness
becomes a sun of giving light for others

Quickly Gaia
will take off her old chains of stone hearts
who don't care about her

We ascend through
her cleansed feathers
joined together to
soar and fly as the angels' wings

We spiral dance
out of the patchwork quilt
of old Earth body
into light threads of her new garment
human weavings blended with
forgiveness and love.

As the Sun and Moon
embrace their sister of void
forming a completion circle
of uncoiled memories

A giving spark returns
out of the dark of Lost Souls
to send a great wave of feeling
through the far holes of our universe

Within a twinkle of God's eye
His beloved lost star,
daughter of man's abuse,
releases Earth cross
stitched in trapped time

She ascends out of the ancient mist
with just a celestial Kiss
This precious planet resurrects
our dream of the power
to be free...

–Meredith Miller

PART II

RESURRECTION IN THE NEW PARADIGM

Meredith
Miller/95

7

Acceptance and Forgiveness

"Acceptance should not be mistaken for a happy stage. It is almost void of feelings...the more (patients) try to deny the more difficult it will be for them to reach this final stage of acceptance with peace and dignity."
-Elisabeth Kubler-Ross

"Lasting change can't happen without acceptance."
-Meredith Miller

"The listeners and scientists are one."
-Stanley McDaniel's Aphorism #100

I am a scientist, a seeker of the truth. When I repeatedly found the truth to be very different from what I had learned and taught about as a scientist, I seemed to enter into an intense grieving process as if I were either to die or to lose a loved one.

The final stage of grieving is acceptance and forgiveness: acceptance of two seemingly contradictory paradigms, one of which appears to be irrationally destructive; and forgiveness for the self and for those who might be unaware of the changes ahead. About one year ago I was at the height of my grieving when the concept of acceptance began to present itself as the only sensible step for me to take in the long run.

Home, Ashland, August 26, 1993

I started to whip up the batter for my blueberry pancakes when author Gary Zukav, our friend and house guest, pulled up a chair at the kitchen counter.

The previous day Gary, Meredith and I had performed a powerful ceremony in the center of the caldera of Wizard Island in the much larger caldera of Crater Lake. The ceremony marked an occasion in which we men were surrendering our egos to a higher will. We brought with us a special sword that sits on Meredith's altar. In the ceremony, Gary and I laid down the sword in acceptance of the wisdom of the feminine, of the Gaia.

For me, there had been good reason for doing this ceremony at this particular time. I had just learned that Mars Observer failed catastrophically, so we would have no high resolution photographs of the enigmatic face and surroundings. Gone would be the long-awaited evidence that artificial structures may exist on Mars, a research effort many of us had been involved in for over a decade. The latest events felt like straws that broke the camel's back of my disappointment in fulfilling my long term commitment to search for life on Mars. In one swift stroke, my foremost lifelong identity as a Mars scientist appeared to go up in smoke. I was in the midst of grieving.

"Gary, as you know, this past year has been bizarre," I began. "This Mars thing is the latest in a series of blocked energy on the search for extraterrestrial intelligence. It's a big mystery that we are getting cut off from learning about the truth. And as you already know, it seems that no matter how hard I try, I seem to keep feeling rebuffed, constantly and consistently, in my attempts to contribute new ideas to the world and to stay afloat in terms of a business as well. The string of setbacks go way beyond probability or chance, and yet it doesn't seem to be that I developed bad breath. My career has always been successful and I've always been sincere about what I do. Do you have any thoughts about this situation?"

He looked at me with understanding. "I would say that you need to make the distinction between the universe supporting you, which always happens, and the universe not supporting what you are doing now. The question is, are you going to continue to attempt to press ahead, already having experienced what you call rejection? It's clear to you as an objective observer that that's not working and the universe isn't supporting it. There's a part of you that doesn't want to let go of that."

"So you keep trying to do it," he continued, "and you keep suffering these disappointments. You're at the place now that no amount of expertise or determination or discussing it with soul brothers is going to change it. That path isn't working for you any more, and that seems evident to all three of us, except that there's emotion involved with letting go of the old path and that's causing you a great deal of difficulty."

"I'd say this would be a fruitful time for you to start an experiment," he said. "Slow down, maybe just for one day. Let go of that path altogether, give it up, let it go...Why don't you give yourself the pleasure of experimenting with a new beginning? It seems to me that the universe is saying that you really want a new beginning."

"What I've been doing," I said, "is that I haven't been accepting these things."

"Precisely. You've been trying to validate them through a system that's much smaller...perhaps you now have the courage to open yourself to a new thought, that the universe is sacred. It's alive, intelligent, compassionate! Parts of you want to say that's nonsense perhaps. You may ask, do I have the courage to trust these inner feelings of mine? That's where the courage comes from, trusting and faith."

I began to recognize that my depression had come from at least two sources. The first was giving up the old paradigm for the new; and the second was the challenge of surviving outside the old paradigm with all its economic and social pressures. I began to realize my experience might be a good case study.

The Importance of Acceptance

My encounter with Gary Zukav over a year ago revealed to me an important aspect of our grieving: acceptance. At that time I intellectually knew what he was talking about, but I somehow felt I had to spend yet another year or two grieving, struggling, judging and resisting. Only now do I seem to be starting to align myself with the universe and look more carefully at the experiment Gary had suggested (which really might take a lifetime, part-time). Only now am I beginning to realize I too have had a limited view of the cosmos prescribed by the box of old science. Perhaps the wisdom of the universe was keeping me away from Mars for reasons yet to be revealed. In any case, I would have to let go and I did. As a result, I started to feel better. My new surrender and feelings of liberation have arrived none too soon.

Because the coming shifts are certain to be very large, we see a great tension between feeling the need to change and the need not to change. In other words, we seem to be inhabiting two contradictory worlds at once. Using free energy and UFO contacts as examples, I have examined in Part I that we must sooner or later publicly embrace these new realities if we are to survive in this world—regardless of the price of change. Zukav believes that the paradigm has already shifted and the universe does not support the old ways based on limited five-sensory reality and the quest for external power.

What arises inevitably from being open to the new information and doing the grieving is that the old cannot dominate our lives for much longer without a monumental struggle. Acceptance and forgiveness represent those steps which free us to do our work. They are the resolution of the apparent paradox of change/no change, the acceptance of death (void) itself and an uncertain after–death experience.

It seems to me that the key to accepting a new reality is a willingness to let go of the struggle, to let go of the de–pression and despair, and to move into a quiet space of accepting our lives in the presence of a dynamically developing new paradigm. We cannot turn back or bargain with the old, yet we can still acknowledge that the old holds sway in the form of continuing denial and suppression. That acceptance is a logical step in our grieving and paves the way toward the transformational phases of creativity, enlight–enment, empowerment, joy and transcendence.

We have seen that the stages of grieving involve unpleasant emotions, ones I can personally attest to having experienced. Based on my experience and that of other new scientists, I am certain that most of those in power who harbor outmoded belief systems will be in for a tough time when they start to move into the grieving process beyond denial. I have encountered many angry men holding leading positions in science and engineering who attend my talks on free energy. At some level they already know the depth of change, and so they back off from the process, perhaps right up until the day of their inevitable transformation, when our civilization will most likely go through some of its greatest moments of change in all of human history.

When that day comes for enough of us, we will all begin to realize our greater potential. We will learn to use our own consciousness to become powerful creators, citizens of the cosmos, eternal beings. We will have seen the mech–anistic, materialistic and deterministic universe as being a mere limiting case of a far greater reality. Science predicts all of this, but my acceptance of it was primarily intellectual and sporadic during my own grieving process.

I have empathy for those who have the courage to grieve. The grief work is very important preparation for transforming ourselves into the new. In the context of the research of Elisabeth Kubler-Ross and others, we have seen that we cannot die with dignity if we have not yet grieved ourselves to the point of acceptance and forgiveness. We cannot take all that old baggage with us, given the mandate for change.

The Process versus the Results

Emotions have a way of feeling all-consuming in time and space. Our fear of experiencing them is keeping most of us in denial because we make the erroneous assumption that we will be angry or depressed for all time. Mistaking the means for the end appears to be keeping many of us stuck in consciousness.

"Whenever a group tries something new," wrote author Marilyn Ferguson, "there is an inevitable period of chaos, and many people, mistaking the turbulent transition for the change itself, decide that they prefer the bad old days. They lose faith and go back to the old."

"When we do that," she continued, "when we run counter to our gut knowing that change must come, we have to rationalize our cowardice. 'Better the devil you know,' we say, 'than the devil you don't know.' And so we cast out the world that might have been."

Acceptance therefore includes being able to distinguish between the chaotic transitional period we are in and what we can become. The development of free energy is a prime example of this unfolding process. Many mainstream people who have not done their grief work will attempt to block the new because of their fear that the system which sustains them will no longer exist. Making such an assumption would be like refusing to get off the airplane that has taken you to a foreign destination. The flight itself is sufficiently familiar but vastly more limited than the destination, yet you would rather stay with what you know about.

The system upon which we are dependent seems to be a house of cards ready to fall. Yet very intelligent, well-meaning people such as my bank president cousin will do everything in their power to keep things propped up in order to avoid a messy transition which could reach to the core of our existing economic and social security. But such actions in the face of the obvious developments on the horizon is like trying to rearrange the deck chairs on a sinking Titanic. The universe can no longer support those actions.

In the context of free energy, this kind of propping up seems to come from playing delaying games in order to optimize profits for the controlling commercial energy sources *a la* J.P. Morgan. Regarding UFO contact, one could look at the Hollywood potential of various encounters while losing the entire meaning of what we can learn from our new reality.

Sooner or later, we will all need to accept our cultural death and rebirth. Whereas Part I focused on how our fear and greed set up a grieving process in the face of radical new developments such as free energy and UFO contact, Part II looks at solutions, on how to weather the storms of change and to create new methods of freeing ourselves from our enslavement to destructive ways.

Part I presented the bad news we need to understand and feel. Part II is the good news of a wonderful renaissance sparked by love and creativity. I will soon be describing new scientific evidence which suggests that if we truly believe we have the potential for a restored Garden of Eden among billions of Edens in the universe, that's what we will get. This is the crux of a Consciousness Revolution, propelled by a new and sacred science. If enough of us accept this evidence, Gary Zukav's idea that the universe is compassionate and supportive of us will happen. As a scientist, I see this optimistic prospect as a working hypothesis for a possible new reality we can help to create in our consciousness. When enough of us do this, it will be so.

Acceptance and forgiveness set the stage for us to formulate our new agendas. They give us an opportunity to see the truth for what it is and to be tolerant of those who do not yet see the truth. They are both mental processes and actions of consciousness that imply a surrender to a higher will. Experiments such as those of the Klingbiels of Spindrift have shown this to be the most powerful force of consciousness that could literally move mountains. Sai Baba appears to understand this and the rest of us may be quick to follow.

Acceptance of the concept that consciousness itself is a major motivating force underlying physical reality, brings us squarely into the core of the meaning of the Consciousness Revolution. But we cannot move into that acceptance without grieving the old.

Acceptance is also a serene feeling of being in harmony with the perfection of the universe exactly the way it is, even though we may not agree with our situation from an egoic point of view. When Meredith and I took on this project, we each had a quiet place inside that knew that. But during our intense grieving, we can sometimes forget our acceptance and often feel swept up in the tides of feeling, of appearing to be the societal victims of the slings and arrows of outrageous fortune. We see this happening to others too—particularly men.

But if we can endure the battering effects of grieving and ride out the storm into the metaphoric void of the donut hole of the hypersphere, we can emerge into a new world relatively immune to the irrational actions of the old world. Being willing to feel our feelings, we may have qualified ourselves to move onto the right side of the table of Gaia—to celebrate our emergence into the new (see painting).

This does not mean we are there all the time. Stressful circumstances sometimes bring us back to the recently felt negative emotions of grieving. The new territory may at first feel unfamiliar to the recent initiate into preparatory depression, the void of death, and then the peaceful (though not necessarily happy) state of acceptance. Accepting and forgiving the struggler inside of us is also an important part of our overall acceptance of the pains of inevitable transition.

Accepting our New Scientific Reality

I accept the realities of free energy, alien contact and consciousness. Having written close to one hundred cautious peer-reviewed papers in the scientific literature that have withstood the test of time, my acceptance of the radical new concepts of this book did not come to me casually. My own assessment has been an intellectual process of looking at experiments and theories that clearly point to their realities.

We do not need a complete theory to explain all of this. Theories will come later; part of acceptance will be our letting go of some of the old models that inhibit us. They simply cannot explain the actions of consciousness. It is encouraging that we do have some models for free energy and other new science concepts already. Some very elegant and exciting ones are described in the concluding chapter. These "straw men" provide a structure which is continually subject to revision as new information comes in. Such is the process of scientific inquiry.

My acceptance did not come to me that easily. I had to plumb levels of my being that were far deeper and more poorly understood than the intellect. I have had to leave the priesthood of science and to grieve the old first. I felt that, in order to understand reality, I had had to travel the world many times to visit and see for myself the experiments and demonstrations performed by those talented inventors, scientists and psychics who are the true pioneers of our time. Along with them, I have often sensed the alienating effects of ridicule, economic deprivation, lack of identity, anger and depression.

Many inventors such as Bruce DePalma, John Hutchison, Sparky Sweet and others have shared with me their own sense of isolation. We are not alone in weathering this process; it is part of being a human on the front line of change.

But it is worth it! When we are dealing with world view changes that are inevitable consequences of observing scientific anomalies, both the intellectual and emotional work needs to be done during the transitional period. Clearing the emotions of grieving provides the ideal environment to move into the creative intellect and intuition that will permit us to formulate those hypotheses and models of reality that can give us the foundation for a truly new science.

The Acceptance of Free Energy and New Science

By its very nature, free energy is democratic, a symbol for getting us out of the electric jail. Part of my own process of liberation is to accept all–that–is and not to spend as much time dwelling on what is not right about the system.

Coming into acceptance also allows us to look at what is good about this crazy contemporary world. Science alone tells us we have concepts that will usher in an explosive renaissance. Moreover, I find in my travels that most people are basically decent, even in the contexts of anthropocentrism, of the adversarial and competitive processes, and of grossly misplaced priorities. In many cases the people are ahead of the Government, the scientists and the media. We have seen, for example, that Gallup Polls show that most Americans believe UFOs are inhabited by alien beings. So time and the truth are on our side. We have a lot to celebrate!

Acceptance is a process that applies to every area of our lives and is the opposite of denial. It might be tempting to deny free energy because of its potential for abuse, analogous to the worldwide abuse of nuclear energy. It is even more tempting to deny both free energy and UFO contact because they threaten our sense of philosophical and economic security. Mustering the courage to accept these things brings on a new sense of responsibility. The practice of our new world view requires that we develop those social structures that will ensure that these things will benefit humankind and Mother Earth. Ultimately we have no choice in the matter.

Acceptance means we need to fully embrace the inevitable in the face of suppression, without fear or greed. The true new patriots are individuals and groups who have compassion for those making the changes, while being directed to achieving extraordinary dreams that may seem now to be beyond reach. Acceptance means grasping an unpopular truth and letting that be the foundation for a new populism that will give us great freedom, a new identity. It therefore means trusting in a divine process to unfold.

Most of all, acceptance provides the fertile emotional, intuitive and intellectual soil from which creativity may spring. It is truly the last supper of Gaia preparing for a resurrection into a whole new existence. This final stage of grieving allows us to die and be reborn both as individuals and as a culture.

Leading metaphysicians have often stated that we humans inhabit a curious zone, one which blends our free will with amnesia about our higher purpose. We have willed ourselves into a struggle and seem to have created some economic monsters. We are going to have to accept all that too, allowing ourselves to feel guilt (bargaining and depression) for our actions. These blocks can be removed by grieving ourselves into acceptance and forgiveness.

When we forgive our actions and those of others, we are ready to heal and transform. It is a peaceful, although not necessarily happy feeling, requiring that we look fondly on all our follies.

For when our true cosmic identity becomes revealed through our inevitable alien contacts, and when we restore respect for our beautiful home planet and start treating it wisely through the appropriate use of free energy, I believe we will be on our way to having heaven on Earth.

8

Creativity

"Where Nature goes to create stars, galaxies, quarks, and leptons, you and I go to create ourselves. The great advantage of this new world view is that it is so immensely creative..."

–Deepak Chopra

"We all look forward to the day when science and religion walk hand in hand through the visible into the invisible."

–Ernest Holmes

I believe that the day we all come together to address this miracle in the void is close at hand. As individuals and as a culture, many of us are beginning to feel a profound inner peace from completing the grieving process, of finally accepting and forgiving the past. As the emotions and mind become still, we have prepared the place for the divine spark of creativity to enter. The first creative steps in new science concern asking questions about reality that may transcend the boundaries of familiar physical perception.

Unfortunately most of the questions scientists and policymakers have been asking are rehashes of the old. The incremental approaches to reform in health care and crime control are symptomatic of the narrow vision of the mainstream scientific and bureaucratic muse nowadays. These attempts have spawned an adversarial Tower of Babel

of special interests that considers solutions that are too little too late. They are only looking at the symptoms rather than the root causes. We will need to be more creative in posing our most fundamental questions and in formulating them as testable hypotheses.

My own excitement about new science relates to the power of our individual and collective consciousness. The mind–over–matter experiments of Jahn and Dunne at Prince–ton, the examination of the power of prayer by the Klingbiels, the numerous successful demonstrations of healing, bio–communications, morphic fields, order from chaos, free energy, UFOs, crop circles, spoon–bending, Sai Baba and Thomaz Green Morton, all pave the way to the reality that our consciousness can play a significant role in altering the material world. Acceptance of the positive results of these experiments and phenomena allows each of us, as new scientists, to become our own experimental subjects. We no longer have to be among the elite, thank God! We won't have to ask the Federal Government for a billion dollars to fund a particle accelerator or space telescope or new weapons system to do our science.

Creativity, Consciousness and Sacred Science

Ironically, the very thing that most interests the new scientist in me—consciousness—is considered largely as a nuisance to mainstream scientists. In *The Second Coming of Science* I gave an example of how some of my own erroneously positive measurements at the telescope of a suspected phenomenon on Venus was probably due to the influence of my wishful thinking on the output of the electronics. Some physicists I know are deeply disturbed by the apparent lack of objectivity in making sufficiently precise measurements of anything which can be influenced by the mind. That includes just about everything and is reminiscent of the ambiguous measurements of the bizarre behavior of particles in quantum mechanics which are inevitably affected by the act of observation.

I define sacred science as a new science that transcends the methodological assumptions of the old science. It uses a new approach (epistemology) that is more than an "objective," repeatable (Cartesian) approach to inquiry in a bigger box. In some cases the older method does seem to work, such as measuring the extraction of free energy using magnetic motors. But in others we may need to factor in our own individual and collective consciousness.

By its very nature, sacred science invites each of us to be our own experiment, to test our hypotheses of reality in our interactions with the physical and nonphysical universe. This approach to science is unitive and therefore must include a whole–systems philosophy that embraces the observer, and cannot be as readily understood solely in the context of definitive experimentation and anecdotes based on the old–paradigm intellectual concepts, such as our exclusive reliance on drugs, surgery, and the materialistic aspects of UFO reports.

Through our experiences as observer–participants, Meredith and I have returned to nature and have spent time contemplating and experimenting with our realities. Much to my surprise, I am verifying for myself the metaphysical hypothesis (as opposed to purely physical explanations) that we are constantly co–creating that reality with the physical universe. Almost daily doses of uncanny synchronicities are exposing me to that reality, and personally point me toward sacred science and the Consciousness Revolution.

The keys to this new paradigm seem to include an opening of the heart, a surrender to a higher will to allow the universe to unfold as it does, and to be aligned to that higher purpose. In other words, our humble acceptance can lead to a divine creative force so powerful that we can resonate with the most important universal messages about our reality. The fundamental experiments on the power of prayer by the Klingbiels and others reinforce this approach of surrendering and listening. Whether or not we choose to believe in the accuracy of these messages, or whatever the source we might ascribe to them (angels, aliens, etc.), many of them seem to ring true. One of the principal observations which seem to be channeled from our alleged wise Pleiadian visitors is that creativity involves a blending of love and information. This could give another clue to practicing sacred science. My own view on channeling is that, if specific predictions can become verified, it is a valid inquiry. Otherwise, it remains to be determined.

I suggest we look at this more comprehensive, sacred science as a blend of "objective" new science and our own personal science, free of the dogmas of traditional science and established religion. I am teaching methods of accessing sacred science, consistent with the more global definitions of science: "way of knowing" and "quest for truth".

Sacred science can provide our needed miracle in the void.

It seems that we can have it all simply be adopting both the personal and the more formal, professional approaches. Of course there will be successes and failures in doing this. We learn from our mistakes and build on our successes. Such is the pioneering nature of true science.

Metaphysics is that branch of philosophy which addresses the nature of our reality. Because it includes the influence of consciousness on matter and energy, it is a much broader vision of the truth than ordinary physics, which normally steers clear of the paradoxes inherent in quantum theory and relativity. As its name implies, metaphysics is a truly more all–encompassing physics, a bigger box which includes concepts from many of the world's leading religious traditions as well as scientific investigations of consciousness itself and other (unabridged) findings of modern science. Aldous Huxley called these basic metaphysical truths the Perennial Philosophy.

Most contemporary Western scientists have given metaphysics a bad rap because of what they perceive as a lack of rigorous proof or difficulty in formulating precise deterministic, mathematical laws. But in denying the evi–dence, these scientists may have thrown the baby out with the bath water. They have painted themselves into a small materialistic corner of the box, a cultural cul–de–sac that inevitably leads to diminishing returns. I know this from years of direct experience: even if Stephen Hawking (or someone else) does come up with a definitive mathematics for a long–awaited Unified Field Theory, his continuing denial of the actions of consciousness could surely invalidate the most general applications of such a theory in an instant of experimentation.

Prigogine's Theory

The scientists' dilemma of defining reality has been recently articulated from within their own ranks. When Nobel laureate Ilya Prigogine reformulated the mathematics of physics he noted that, in the most general case, the equations lack the time symmetries inherent in quantum mechanics and classical dynamics. He sees chaos, or creativity, as the wellsprings for new laws of nature that transcend Big Bang materialistic determinism.

"Quantum mechanics is not final," he said. "You have to find a larger theory for unstable systems that contain irreversibility." Life forms represent one such system.

Prigogine is the highly respected author of chaos theory, which, as I explain in *The Second Coming of Science*, may be an important aspect of understanding the physics of free energy. His latest work will almost certainly be part of the foundation of a new sacred science around which many of us could begin to do fruitful work. His new model has received the following favorable comments from eminent scientists:

Rupert Sheldrake, cell biologist: "Prigogine's work is a major contribution to the new view of nature that will underlie the science of the next century—the view of the cosmos as creative, evolutionary and alive...New laws appear as nature evolves."

Larry Dossey, consultant to the Office of Alternative Medicine, NIH: "(Prigogine) has shown how chaos can give rise to order, complexity and beauty—all in a world that classical physicists said is destined to entropic messiness, disorganization, decay and death."

Keith Harary, Institute for Advanced Psychology: "Chaos or creativity, whatever we call it, it seems clear that the process of transformation of existing "laws" is a fundamental force in nature...Prigogine has contributed significantly to our disavowal of the arrogant notion that science as currently construed will reduce the workings of the universe to predictable equations...The human element—the development of art, culture, society, even science itself—may exert a more basic influence than we dare imagine."

Willis Harman, president of the Institute of Noetic Sciences: "Prigogine is rightly concerned with the social consequences of the predictability dogma. Perhaps we are, as he says, 'at the beginning of science'."

Metaphysical versus Physical Hypotheses of Reality

During the transitional period in which we find ourselves having difficulty in modeling our reality, scientists will find it useful to formulate alternative hypotheses based on a fundamental question. One example would be the hypothesis that the universe is interconnected by resonances that transcend time and space. This is a metaphysical hypothesis which, if shown to be experimentally true, would contradict the hypothesis based on a traditional physics that would not allow such resonances to occur because of rigidly-defined cause-and-effect relationships that depend on five-sensory perception.

I believe that the main task of sacred science is to test physical and metaphysical hypotheses to determine the nature of reality. Sacred science explicitly probes and tests metaphysical hypotheses while the old science generally excludes non-physical reality.

A hypothesis is merely an educated guess about reality that can be tested by exploratory experiments. Competing hypotheses need not always be mutually exclusive and are subject to refinement. As we perform the experiments, we can begin to express some models, theories and refined hypotheses to probe further. Princeton scientists Robert Jahn and Brenda Dunne call this process the "two-step" between experiment and theory, which would lead to increasingly mature developments of the new science.

In the examples I cited, the findings of Prigogine, Sheldrake, David Bohm and others clearly support the notion of interactive resonances. These cannot be explained by the mainstream physics of the Big Bang/Unified Field theorists nor by the quantum theorists. This validation of a broader metaphysical hypothesis then leads to a new theory of an evolving, conscious universe whose resonances can even change our cherished physical laws, which we have assumed to be immutable.

Part of a new scientist's creative action is to formulate those hypotheses which can cover a wide gamut of topics, chosen initially at a most fundamental level, such as, "We are eternal" (metaphysical hypothesis) versus "Death is the end of our awareness" (physical hypothesis); "We are mystical beings enshrouded in a mystery which can be unraveled by clues and symbols of nature" versus "All we are is what we see, hear, feel, smell and taste, and expressions of nature are random, outside events"; "Holistic health provides a wide range of effective healing options" versus "Only Western medicine has the means to heal through pharmaceuticals or surgery"; "The vacuum of space contains a dynamic ether, or zero-point field, whose energy potential is virtually unlimited" versus "The vacuum of space is empty and without consequence"; and so forth.

In summary, the major creative force in science lies in asking the deepest questions, formulating alternative hypotheses, devising experiments to test them out, reformulating these hypotheses, and eventually coming up with a theory or model. I believe that the collective efforts of Prigogine, Sheldrake, Puthoff, Jahn, Spindrift, Backster and others have gotten to such a point of refinement that it is possible now to pose theories that transcend quantum mechanics. The phenomena of free energy, psychokinesis, spontaneous healing, morphic resonance and bio-communications all inevitably point to a new science that will liberate ourselves from the box (see Glossary for definitions of new science terms). There is probably no better test of these theories than our own personal experiences of meaningful coincidences (synchronicities).

Exploring Personal Synchronicities

A few weeks ago, while I was staying at a hotel in New Jersey, I dreamed about being a spirit guiding a man to swing his golf iron to make a perfect hole–in–one from about a hundred yards away. Shortly after I woke up from the dream I turned on CNN to get the latest news on the collision of Comet Shoemaker–Levy with Jupiter. While an ad was playing I scanned the dial out of an innate curiosity, and was very drawn to the live coverage on another channel of the final round of the British Open Golf Tournament.

As I picked up on the action, the Tournament leader was already entering the clubhouse with a one–stroke edge over runner–up Nick Price, who in a minute or two was about to attempt a forty–foot putt that would give him an eagle (two strokes under par) on the seventeenth hole. Not knowing why, I felt drawn to looking at the golf tournament even though I rarely watch golf on television, don't play golf, and have never recalled any dreams about golf. In fact I could not understand why I, as a former professional planetary astronomer, once a colleague of the comet discoverer Gene Shoemaker, and eager to be up–to–date on this most exciting and rare celestial event, preferred to watch golf! It made no sense but I answered my intuition and stayed on the sports channel.

Within five minutes of turning on the television and one half hour after the dream itself, Nick Price sunk his long putt, and after parring the eighteenth hole, won the tournament. The metaphysical hypothesis would posit that the dream was precognitive, that I experienced a resonance of a meaningful coincidence of events. Precognition is a commonly experienced and measured phenomenon, reported by Jahn and others in controlled experiments.

Because I had not been trained to look at these things or to believe in their significance, I at first denied any meaningful aspects of my golfing vision. Also, having been recently bombarded by unusual coincidences, it was becoming more of a daily experience and therefore not all that remarkable.

Only later did I begin to see the deeper meanings behind the pairing of events. The great psychologist Carl Jung called these resonances synchronicities—physically noncausal events that happen close in time that transcend random coincidence. The classic example of a synchronicity is when we think about somebody and they call. We all have these experiences, and they beg an explanation mainstream science cannot supply.

In his best-selling novel *The Celestine Prophecy*, James Redfield emphasizes the importance of meaningful coincidences. The first prophecy in the Celestine manuscript states that the acceleration of our experience of synchronicity presages planetary transformation.

Speaking from personal experience, Meredith and I have noted hundreds of these resonances and have recorded them in taped and written diaries. Perhaps our most dramatic synchronicity was how we became partners in life. I had invited Meredith up from southern California as a casual friend to house-sit at my Oregon home while I was in India during January 1992. Surprisingly enough, she came! But not for reasons I would have expected: she used the house as a base from which she would get together with a handsome, much younger man, a massage therapist from Seattle. I too had been attracted to Ashland by a pretty, much younger woman. Neither of our flings worked, but Meredith loved the house, we soon fell in love, and she moved up here some months later. We then bought the house, and in independent actions at different times, each of our mothers voluntarily donated exactly the same amount of money (to within one per cent, adjusted by interest) to help us secure the house.

There is even more synchronicity to these events. Meredith had grown up on the nearby Rogue River, a magical childhood in the woods which had inspired her to become an artist. She was beckoned to come back here. Also, immediately after my India trip I had some pre-booked talks in Phoenix and San Diego. Meredith lived in San Diego and had business in Phoenix exactly when I was there. This brought us into enough intimacy to stir up interest. The painting Meredith did at my home was delivered to a magazine in Phoenix, all in my presence in Phoenix. The painting showed a couple coming together in joyous passion and was given the name *Coming Home* by the magazine, not by Meredith. Little did we know how prophetic all this would be.

The two of us had had similar childhoods in nature, talking to inanimate objects and developing our imaginations beyond cultural norms. Our unpredictable, fluctuating incomes were (and still are) almost exactly the same. Even our press kit covers, independently created years ago, were almost identical—glossy white with a large purple signature at the bottom. The current writing and painting project and the prospects of our moving to Maui (a long-held dream we've each had for decades) are other signs we are two creative peas in the same pod, best friends moving through our karmic journeys as if some part of us had done all this before. We have also discovered some negative patterns borrowed from our parents which seem to be most dramatically set up to compel us to learn, grow and transcend.

It is not my purpose to list scores of other synchronicities or to try to convince you they all transcend the possibility of randomness posed by a mundane physical hypothesis. But the sheer number of our synchronistic experiences defies any common-sense physical paradigm. It becomes a valid inquiry in the context of sacred science.

There is much more richness in all of this than noting mere coincidences in timing. More often than not, synchronicities seem to represent symbols or clues to our greater reality. For example, the golf scenario drew my attention to a recent perception that I was right on track, that my heavy grief was appropriate. I resonated with the feeling of being on target, as I moved through the void, bull's-eye, into the tiny donut hole of surrender and acceptance.

Later that night, coming home from the airport to our Oregon house, Meredith and I nearly struck two vulnerable fawns standing in the middle of the highway. We just stopped short and had a one-minute-long mystical exchange with the fawns. Borrowing from the tradition of native Americans, we looked up the meaning of the fawn as a medicine animal. We found it symbolizes a gentleness with no fear, an acceptance of whatever circumstances may prevail. Everything would be O.K.

The meaning of the encounter to us was profound: we were both ready to accept whatever presented itself to us, no matter how threatening it may seem. Our grieving may be nearly over, except for those nasty residual emotions that tend to hang on for awhile. Faith, trust and courage would get us through. We needed these little deer to tell us we were right on target and there was nothing to worry about, just as Sai Baba had told me. A few days later, quite unexpectedly, we completed a land sale that had been riddled with uncertainty. In a few strokes of the pen, the transaction relieved us of two years of financial challenge—thanks to a gift from Gaia heralded by some powerful synchronicities that had profound meaning for us.

To my way of thinking, recording these experiences and relating them to what is happening to us at a deeper level, constitute sacred science at its best. It also has predictive qualities like any other theory. For me to transcend the notion that we are all on a mindless, random walk in an entropic universe is a great awakening. It is our miracle in the void, freeing our energies to align with the universe the way it truly is—not the egotistical, limited physical way we scientists had thought was the whole truth.

Almost daily, Meredith and I encounter animals, hear sounds and rustles in nature, observe aspects of the sun, moon and planets, or watch a Star Trek episode. We began to notice that this sacred dance of nature and humanity resonates with our feelings and with those heartfelt personal or global issues we were addressing at the time. For the most part, we have let go of the trappings of the city and have entered the realms of understanding the interrelatedness of all things in nature. But even the cities are full of synchronistic symbols such as relating your lives with dreams, billboards, license plates, magazines, telephone calls, television and radio programs.

When we see doubles close in time (e.g., the two fawns, two rattlesnakes, two rainbows) we sense the meaning to be even more profound, and that usually turns out to be so. Nature seems to be a reflector of a deeply felt inner experience of truth. It is also an "entity which speaks to us if only we were to learn its language", as Meredith has often pointed out to me.

There is so much richness to all of this. It is ironic that we Westerners, who feel so smug in our ability to measure and control what we call objective reality, can really penetrate only to a very shallow depth of interpreting its true meaning. Synchronicities provide an interesting structure for storytelling or myth-making, which helps us unravel the mysteries of our lives as well as newly emerging consensus realities that seem to have little to do with the old paradigm.

During those times I seem to struggle, and in visiting bustling cities, the resonances sometimes vanish for me, perhaps in part because I do not notice them as much.

Meredith is a sensitive artist who has looked at these things for most of her life. She is my teacher to take notice. I have been at this for two years now. The sheer volume and uncanny unlikeliness of these events and what they reveal in terms of deeper meaning, convince me there is something to all of this. From taking polls of my audiences, it seems that the first Celestine prophecy is being fulfilled for many of us. The seeming acceleration of synchronicities (or our awareness of them) further challenge the dying notions of the old science. What meaningful coincidences have you noticed recently? Under what circumstances do they become most apparent and how do they resonate with what seems to be going on in your life?

Another important Celestine prophecy is the importance of understanding the broad sweep of history rather than through the filter of our contemporary industrialized and sanitized twentieth century version of it. Such a perspective reveals much of the lost magic of the aboriginal cultures and further reinforces the validity of the metaphysical hypotheses of new science. The argument becomes even stronger that we will sooner or later need to abandon the materialistic, anthropocentric world view. We will more clearly see our provincial viewpoint as an anomaly of the past, an activity which is no longer aligned with a universe created anew. Meanwhile, it can often be challenging to insulate ourselves from the grips of the old world view, of our own separation from ourselves and from the universe, whether it be environmental or personal pollution.

The Divine Spark

As small children, Meredith and I were close to nature and regarded it with awe, reverence and wonder. I had also envisioned exciting trips to the Moon and to Mars. When I was very young I often had a fantasy about being the World's Secret, that I was watched by others who put me here (God? aliens?) about what I would be doing here and how I would carry through on my mission. Later that notion seemed too egocentric—after all, who would care about only one of several billion people here? The fantasy evaporated as I descended into the box of our Earthly being, the electric jail. The mystics' veil of Maya illusion had dropped. From my "mature" adult perspective, little did I know there might be some truth to that fantasy in the presence of a cultural conditioning that took away the magic. Yet the fantasy was almost certainly a pure experience of consciousness, consistent with my re-enlightenment to the interactive universe.

I now perceive the wisdom of my childhood perception which for decades was brainwashed out of me. It relates to the realities of alien contact, divine intelligence, and the oneness of all creation. Somehow I had accessed a spark.

My metaphysical hypothesis is that we all have this divine spark. We can all use our forces of consciousness to transcend our older levels of perception based on objective, three-dimensional, materialistic science. We all have authentic power (Gary Zukav's phrase) within us that has been covered over and over again until we put ourselves into electric fences that are so confining that our old culture has literally choked itself off from most of its transcendent potential. The divine spark has been emasculated by a competitive economy in which people seek external power and acknowledge only physical reality. The universe will not permit this approach to go on for much longer, as Gary Zukav has so eloquently pointed out in *The Seat of the Soul*.

That divine spark within each of us allows our authentic power to emerge in the same magical ways as the experience of a child. It is a subtle and creative energy that brings on our transcendent truth and provides the right metaphor for free energy. Free energy then becomes an action of consciousness from within. It can eventually heal, transform, transmute and enlighten all of us. Sacred science can guide us to bring on a free energy that loves the Earth and all creation.

Many free energy inventors have repeatedly told me their inspirations came from a mystical insight, vision or experience. This is their divine spark which promises to enlighten the world. For nearly one hundred years their work has been suppressed. Let's now give them a chance, let's give all of us a chance to embrace this powerful new science of love and creativity from the heart.

Embracing free energy, UFOs and the experiences of consciousness, creativity, synchronicity, and childhood fantasies all point to our profound interconnectedness. They are the way-showers toward embracing a divine plan through the practice of a sacred science. Orchestrating the big picture behind the scenes, we can begin to perceive an elegant cosmic choreographer of such wealth and wisdom to have provided these opportunities. We cannot help but surrender in awe. When we start to do that, the universe changes and we change.

The news is indeed very good. Magic is real!

"...The entire vista of science takes on a new dimension once we understand that matter and life, energy and consciousness are created from within. Then we begin to see how man can work not in fear of natural forces, but in true harmony with nature."

–John Davidson in The Secret of the Creative Vacuum

9

Enlightenment

"As a man who has devoted his whole life to the most clear-headed science, to the study of matter, I can tell you as the result of my research about the atoms, this much: There is no matter as such! All matter originates and exists only by virtue of a force which brings the particles of an atom to vibration and holds this most minute solar system of the atom together...We must assume behind this force the existence of a conscious and intelligent Mind. This Mind is the matrix of all matter."

–Max Planck, lecture given in Florence and cited by Davidson

Enlightenment can take place when individual scientists verify for themselves, at both a mental and emotional level, a countercultural metaphysical reality. It is the moment of discovery after the hard work of searching and grieving and creating. It is the "Eureka" of Archimedes when he discovered the principles of buoyancy, and of the forty-niners in America when they discovered gold. We are like alchemists truly on the threshold of discovering a new, much more rewarding gold in the void.

No amount of skepticism or lack of support or cultural aberration can undo those special moments of enlightenment. When enough scientists become enlightened and can provide mutual support to move ahead with their ideas, the foundation for our awareness of the paradigm shift is laid. Then we can become empowered to help change the world.

I have discovered that more individuals than we can imagine have become quietly enlightened with alien contact, free energy or holistic healing. Even more of us have felt the creative sparks in those directions but are still looking for sufficient consensus and/or inner drive that can lead to experiments or insights which are enlightening.

As a mainstream scientist, I had felt enlightened by many experiences of discovery within the relatively narrow context of a mechanistic science. Since that time, my meta–physical experiences had once seemed to appear only as disjointed anomalies that revealed little or no deeper meaning. They are now becoming the core themes and hypotheses of life.

As I evolve, my science also evolves. As nature evolves, its laws also evolve. These are the main hypotheses of Prigogine, Sheldrake and other pioneers of science. And we ourselves may be able to take part in defining and formulating these laws, the ultimate expression of which is that we all create our own realities. Our world appears to be a five–sensory and material one. But, at a more fundamental level, it is empowered and conscious.

Alas, these possibilities seem remote while we are denying, grieving, holding onto the old. One needs to be per–sistent in this lonely path toward enlightenment. Yet, because of the ecological mandate we now have, that path needs to concern all of us, not only the individual who might become illuminated by sitting in a cave in Tibet eating bean sprouts.

While at the time I didn't realize it, the ceremony at Crater Lake with Meredith and Gary Zukav seemed to have helped create enough of an atmosphere of acceptance, creativity and enlightenment, to make things move. Just three weeks later a number of synchronistic events, apparently designed by the universe, presented me with my first major experience of enlightenment for quite some time.

Rome, Italy, September 13, 1993

As I was about to walk onto the stage set of the Maurizio Costanzo Show, I still couldn't believe I was so suddenly in Rome. Just five days earlier in our Oregon home, I had had an extraordinary coincidence in the timing of events that triggered this trip. All within five minutes, the invitation to appear on the television program had come to me, a piano tuner had broken a D–string of my piano in a bizarre accident, an old sweetheart named Dee had called for the first time in over a year, and a long–awaited van had been offered to us for an important art show in Southern California. After a year of no action, the universe had suddenly presented to me an agenda worth looking at. That evening, Meredith and I had headed down the hill to hear author Marlo Morgan speak of her spontaneous walkabout and extraordinary adventure with a group of Australian aborigines, reported in her best–seller *Mutant Message Down Under*. Her inspiring talk had kicked off my own walkabout into a unplanned and uncertain future in Rome, of all places...

And so that next morning I had been off to New York state to give a workshop and then to Rome, flying the red eye for the videotaping that next day. The main subject of this particular show, the most popular talk show in Italy, related to the recent catastrophic failure of Mars Observer and the possible NASA cover–up on photographing the face on Mars. I came prepared with the latest copy of *The McDaniel Report* to discuss NASA's reluctance to look at the evidence.

I began to realize that the void I had entered in dropping out of my Mars work and doing the Crater Lake ceremony may have created this opportunity to travel to Italy to talk about Mars. If Americans were blind to the insights many of us outside scientists had about Mars, the Italians might not be. At least energy was beginning to move. The D–string accident brought my mind back to a trip to Rome I had once taken with Dee; that trip had been full of challenges. On another trip to Rome I had met with the Pope, a symbol of my Catholic upbringing. During yet another trip as a twenty–three–year–old, I had rented a Lambretta that I dented and drove in a dramatically futile attempt to pursue a girl I never caught up with. I have had much history with the Eternal City, perhaps even more than I might have imagined.

The television show itself did not go as I thought it would. We were on the set for three hours, and it was only after two hours of show–biz distractions that I had a chance to talk. Sitting poised with an earphone in one ear and an Italian–to–English translator whispering into my other ear, my time came when Costanzo and some other guests began to grill me about whether the Mars Orbiter's alleged failure might really be a cover–up story for NASA's obtaining photographs in secrecy. Not being an insider, I did not know the answer to that specific hot–off–the–press question. Yet that seemed to be the whole reason they flew me to Italy on such short notice! I believe I might have disappointed them with no sensational news.

Such is the fickle nature of the media. Actually I did report to the audience the strong evidence for the artificiality of the Martian features. I cited McDaniel's work showing instances of ridicule and propaganda issued through the years by NASA, Sagan and others. I also talked about the UFO abduction phenomenon and was able to draw out one articulate woman in the studio audience, the wife of a doctor from Sicily, to tell her story of an alien encounter in her home.

And then, Roman circus style, almost the entire guest panel of about six old–paradigm Europeans (including a NASA–funded geologist from the University of Rome, a philosopher of science from the University of Vienna, a leading Italian journalist, Costanzo himself and some of the others) began to probe me, labeling me as a "Martian". The professional geologist dismissed the face on Mars as being a mere trick of lighting and shadow, analogous to natural facelike rocks on Earth. This kind of *ad hoc* debunkery represented NASA's party line and sounded impressive. Yet it all seemed to be so naive. More ridicule took over the show, but the Sicilian lady and I held our own.

After the show, I withdrew to the hotel, having made few new friends. I luxuriated myself into depression for two days before my flight back to the States. At least I had the enlightenment of knowing more about my "enemies" in thought. They were intellectual pundits cynically dismissing expressions of profound new concepts which violate a conventional wisdom safely akin to their own vested interests. I found no adversaries with that much to say, except for the same old *a priori* skepticism which was as unscientific as you could get.

Rome, Italy, September 15, 1993

I entered the Vatican Museum just as it opened and sprinted toward the restored Sistine Chapel so that I could have it all to myself for a few moments. Suddenly my intuition told me to look outside a window along the corridor.

Down to the right I saw a courtyard which I felt drawn to look at for awhile, as others headed past me toward the famous ceiling fresco painted by Michaelangelo.

The courtyard looked very familiar to me. I began recalling a hypnotic regression I once had (as reported in

Exploring Inner and Outer Space) about an alleged former life in which I was burned at the stake or stoned to death for heresy. I realized that what I was looking at was the same courtyard I had visualized during the regression. My Catholic upbringing, with its associated guilt feelings, added impact to the intensity of my Vatican visit. Shivers went through my spine and I felt mildly nauseous. Could it be, I actually experienced that lifetime? Was my modern persecution in front of millions of Italians but a replay of earlier experiences on this hauntingly familiar land?

I then visited the Forum and Coliseum. Before returning to the hotel I felt in touch with the even earlier times of the Roman Empire. Taking advantage of the five-star ambiance offered to me at the hotel, I sat alone in the restaurant wining and dining myself to excess while reflecting on my experience. I observed apparently wealthy business-men also indulging, and I felt this curious mixture of guilt about my decadence and persecution about my heretical ideas.

Only seven months later, on another sponsored trip to Italy with Meredith, we both seemed to harbor the same schizophrenia between being indulgent patricians and humble Franciscan monks. Could it be, we experienced at some level both lives? Many newer Italian synchronicities, added to my earlier visit, seemed to be set-ups from which we could learn about what to do about some old energies that seemed to haunt us.

One very important metaphysical hypothesis, based in part on Hindu and other Eastern religious practice, is that we are all souls who have reincarnated to balance out a natural law called karma. In Italy I felt an eerie resonance with the behaviors of leading scientists, politicians, clergy and bureaucrats spanning at least three different historical periods: the Roman empire, the Inquisition/early Renaissance, and the modern era. Each epoch has marked times of the abuse of power, ignorance, complacence and arrogance among the ruling elite.

My enlightenment about this comes from the richness of the Italian trips which present a powerful myth for Meredith and me, based on patterns that intersect with historic times and places that are significant for us and appear to keep bringing us back there. We visualized that a cosmic choreographer placed us there to balance an energy intimately related to unresolved issues from the past so that we could clear it. Powerful emotions were felt each time. Maybe these encounters would better qualify us to become more empowered to assist in the coming Earth changes. The scientific evidence for reincarnation is virtually undisputed in at least 2000 cases of children surveyed by University of Virginia professor of psychiatry Ian Stevenson, as described in *Exploring Inner and Outer Space*.

Our myths takes on even deeper meaning when we began to realize that our effectiveness in doing our work in the world might be too compromised by slipping into former roles either as aristocrats steeped in the system or as martyrs resisting it. When our acceptance and our enlightenment takes precedence over Roman or Catholic or contemporary anthropocentrism, we may begin to feel more like transcending the mundane and embracing the divine. The mysteries of our lives begin to unravel, even without having to take the notions of karma and reincarnation beyond the metaphysical hypothesis stage—sacred science at its best.

Looking over my resonances in Rome, I quickly learned that regardless of what we might or might not believe to be true, we can all be enlightened by the extraordinary dramas of our lives to the degree we can surrender to the will of the universe and see what it presents to us. In emptying out, I discovered that the truth can be stranger than fiction, that the universe is conscious and compassionate in presenting us with opportunities to grow and experience. All we need to do is to show up and jump in. That becomes the miracle in the void. What might be some of your miracles in the void?

Why We Must Take Risks Now

Life is an adventure so what is there to lose by fully embracing it? In a Star Trek episode, the diabolical and powerful character "Q" gives Captain Jean Luc Picard an opportunity at the moment of his impending death, to re-choreograph his life in a more cautious way, thus nullifying the cause of his death (a failed artificial heart). Then he could live. In the safer life, Picard became a humble astrophysicist and junior officer on board the *Enterprise*. In the end, he much preferred his real, more gutsy life as a captain. In selecting his own life back, he ended up living after all.

As a culture, we too are on our death beds as we see our fragile spaceship Earth disintegrate. The only way for us to make a difference is to take the same kinds of risks Picard did. Unlike the good captain, we probably won't have to get into challenging physical fights in order to make a difference. Mahatma Gandhi and Martin Luther King have shown that there are other ways.

Gaia is also on her death bed, preparing a Last Supper for us to partake before her resurrection—*our* resurrection. If we can but trust that this will happen, then it probably will. Meredith inspired me to look at the art of sacred science, the art of predicting a societal miracle in the void. She taught me that synchronicities and dreams and products of the imagination were clues to deeper meanings. Those meanings unfold like magic stories, myths whose richness can exceed our wildest fantasies.

Once we begin to accept our unfolding personal and cultural myths as nonphysical aspects of reality, we can understand the hero's exciting journey. It becomes a treasure hunt, a quest for the Holy Grail, bread crumbs to follow on the path to our transcendence. We all have the opportunity to follow that path of surrender to Nature's will. By accepting, creating and cooperating with the universe, we can prepare ourselves for the big shifts on the planet, through our own enlightenment. We can once again get glimpses of the unfoldment of a higher order, of a divine plan. Some call that higher order God.

For me, the quest for understanding that higher order began to be answered in a most unlikely place: Japan. Here is where the practical, commercial aspects of life most effectively intersect our enlightened dreams. In the Land of the Rising Sun, there is a spirit of openness, generosity and abundance, curiously blended with an environment that appears as a polluted beehive. The Japanese could surprise us once again. Being an oil importing nation, they have every reason to develop commercial free energy; and that is exactly what they are doing, free of the vested political and economic interests that have inhibited us here in the West. During a recent trip, I found some enlightenment there.

Tokyo, Japan, stopover en route to India, Feb. 10, 1994

Shiuji Inomata and I got into a taxi at Narita Airport for a $150 ride to his laboratory in the Japanese Space City of Tsukuba. In a style Japanese hosts are famous for, he had gone out of his way to pick me up, take me to dinner, demonstrate his free energy device, and send me back to the airport in another equally expensive taxi—all at his expense.

Inomata is a senior scientist of 35 years' tenure with the Japanese Government, at the Tsukuba Electrotechnical Laboratory. He has a Ph.D., is author of dozens of peer–reviewed papers, and is president of the Japan Psychotronics

Institute. He is not the stereotype of the Saturday garage inventor. He has built a version of Bruce DePalma's unipolar generator (sometimes called the "N-machine", connoting the production of power to the Nth degree). Our conversation was animated during the two hours in the taxi, which traversed the hodgepodge of roads, powerlines, exotic billboards and an over-industrialized landscape. We had a pleasant sushi dinner at a local establishment. After nightfall he took me up to his laboratory to demonstrate his device for me, a tabletop machine that showed anomalous outputs once it ran beyond a certain threshold of speed.

He spoke to me enthusiastically about a seminar he had just given before an overflow audience of six hundred leading industrialists, academics and governmental scientists. "I believe the people of Japan will support free energy," he said in mostly understandable English with an endearing grin. "We have no domestic energy supply. We need it and we're not suppressing it. Toshiba will spend two million dollars on superconducting magnets for new unipolar generators (N-machines) that will give us over-unity, more electrical output than input. Then we can manufacture them...Yes!" We nodded together in a manner exceeding mere courtesy.

I knew what he was talking about. While most American decision-makers were "floundering in the backwash" (a favorite John F. Kennedy quote), denying and suppressing free energy, the Japanese were organizing themselves to take off. Inomata and several other enlightened and highly qualified scientists are moving into this area, and almost certainly it will be only a matter of time before we see Japanese free energy devices dominating the marketplace.

Inomata's demonstration of his machine, a simple small-scale replica of the invention of Bruce DePalma, an expatriated American, convinced me this very intelligent and well-credentialed scientist was on his way to success in concert with Government, universities, and industry. Does this story ring a familiar bell? When will we ever learn?

Inomata then showed me a diagram that immediately made sense to me in my own path to enlightenment. It was a triangle whose apexes represent mass, energy, and consciousness. This one gestalt was all I needed to synthesize in my mind the very simple fact that our current science considers only the internal and external relationships between mass and energy (immortalized by Einstein's $E=Mc^2$), while totally ignoring the influences of consciousness (creativity). The addition of the relationships of consciousness with itself and with mass and energy adds a new dimension to the linear mass–energy paradigm, and opens new doorways to understanding our greater reality. The following diagram embodies the essence of our paradigm shift in science.

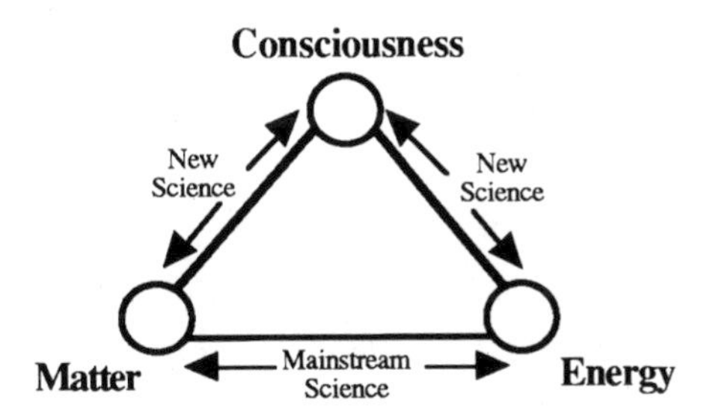

I then began to think about how all this might relate to new science, that each apex and line of the triangle reveals profound new aspects of our truth. For example, free energy can either come out of our consciousness or by designing a device made of matter (consciousness and mass create energy). Alien contact, materializations, psychokinesis and holistic healing can happen in the real physical world when either consciousness or energy acts on our bodies (consciousness and energy influence mass—our bodies).

The domains of consciousness itself are just beginning to be understood. When I look at the apexes from a new science point of view, I see a fourth apex which could represent death, transformation and rebirth. This added dimension then creates a tetrahedron.

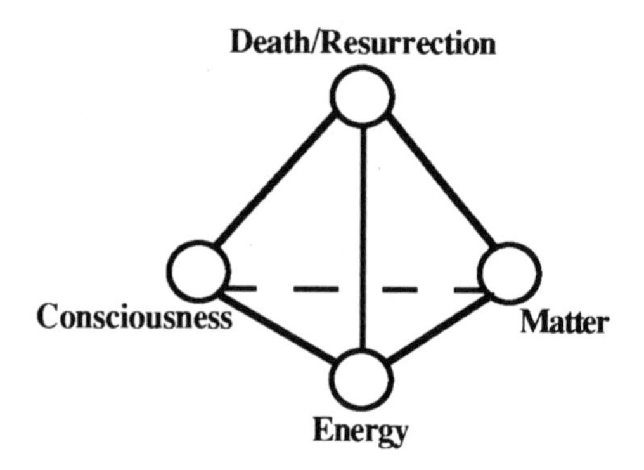

In considering research on free energy, alien contact, consciousness, mind–over–matter, healing, near–death and after–death experiences, spirit communication and possession and reincarnation, we seem to have reenergized a multi–dimensional sacred science that considers the interaction of at least four major elements of our reality. I began to develop a model that made more sense to me than mere materialistic science. And, according to evidence I will present in the last chapter, the concept of mass itself may have to be discarded—which leaves energy, consciousness, life/death as the most important parameters.

These experiences of personal breakthrough in the presence of my Japanese colleague were perfect preparation for the first demonstration of an over–unity (over 100% electrical efficiency) free energy device I had observed. The scientist this time, also an eminent and enlightened being, comes from India. His name is Paramahamsa Tewari.

Karwar, India, February 12, 1994

The orange and crimson sunset over the Arabian Sea provided an ideal backdrop to my welcome arrival at the guest house of the administrative headquarters of India's largest nuclear power plant construction project now underway, about thirty miles upstream near the village of Kaiga. Chief project engineer Paramahamsa Tewari pulled up in front of the guest house in his comfortable government car.

"Hi, Brian, it's so good to see you!" he said as he looked at me with his enthusiastic childlike brilliance. We warmly hugged. We had met each other once before in Colorado at the First International Symposium on New Energy. We had felt instantly bonded as soul brothers and had discovered we were mutual friends with Sai Baba. I had taken him up on his generous invitation to see his free energy machine in action.

Getting to Karwar was quite an experience, but certainly worth it for the opportunity to observe the only visitable free energy generator in the world known to reliably and repeatedly produce over–unity electrical efficiency. Very few scientists have had the opportunity to travel to this most remote and remarkable site on Earth. Karwar is about as far as you could get from America.

The obstacles in my journey felt formidable. I had been paged at the airport in Bombay just prior to my flight to Goa that the road onward to Karwar had been blocked by local violence between competing religious and political factions. On his way to the airport to pick me up, Tewari's diligent and amicable administrator, Mr. Menon, still braved the road block with multiple rides in taxis and bicycles across fields to greet me on time and to deliver me to my rendezvous with Tewari three hours later. During the hot bumpy ride, I looked at all the poverty and environmental degradation that has so gripped this subcontinent—even in the relatively lush southern coastal region I was passing through.

"Come, let's go," as Tewari led me into the laboratory with his entourage of assistants. In his own setting, he seemed to possess a curious mixture of acting the part of an important governmental official with being the humble, connected and spiritual man I knew him to be. Only in India, I thought. In spite of all the obstacles in getting there and my jet lag, I was excited at the prospect of seeing his demonstrations.

While I was dashing around the lab taking photographs and noting one of the most amazingly colorful twilights I have ever seen, Tewari started up one of his generators, a large noisy clunker weighing a few tons (see photos). When the machine revved up to about three thousand revolutions per minute, his meters began to show efficiencies of over–unity. Unless this man of obviously high integrity were shown to be a fraud, a highly unlikely prospect, I was observing a true free energy machine in operation.

Like Inomata, Tewari built a unipolar generator (N–machine) consisting primarily of two corotating magnetic disks. He calls his two free–energy devices space power generators. One operates on direct current and the other, alternating current. Both generators typically produce electrical outputs that exceed the inputs with efficiencies ranging from about 200 to 400 per cent.

Also like Inomata, Tewari is a highly credentialed scientist with a Ph.D. and with several publications to his credit. He had also won first prize for his N–machine in an international science competition. His position with the Indian Government is about as high as you could get outside of the remote capital city of Delhi. He has thousands of employees working for him on a project whose gigawatt potential represents an investment of over a billion dollars. Ironically, nuclear power development in Third World countries such as India also present a considerable safety hazard—one that could be relieved by the project director's own space power generator.

"Commercial space power is now possible," Tewari yelled at me over the unmuffled sound of the generator. "I have now solved the problem of back torque so we can now have unlimited electricity for humanity."

He shared with me his own story of discovery and enlightenment in his laboratory notes of the previous December 26th, when he was able to engineer himself around a problem that beleaguers researchers of magnetic free energy devices: it seems that as free energy begins to come off the magnets, most all of the electricity the generator produces backs up on itself to counteract the sought-after effects. Tewari enthusiastically described to me his own "Eureka" to get around the problem.

The overheating of his motors necessitates shut-down after about five minutes of operation, so Tewari's existing machines are not yet practical energy sources. But they do clearly demonstrate proof-of-concept, and he appears convinced he has overcome his problems for the new model he is building, which could produce kilowatts and eventually megawatts of power.

Nobody believes that India, of all countries, could begin to mass-produce free energy machines for export. But in today's strange world, it may be possible, especially with a supportive international effort. Paramahamsa Tewari has already shown the impossible to be possible.

The rest of my trip to India added icing to the cake of enlightenment. I flew home after my encounter with Sai

Baba, in which I realized we can create matter and energy out of nothing. Therefore nothing is something. On my approach to San Francisco after the long journey from Bombay and Tokyo, lightning struck the airplane in a shocking wham, just as the jet was passing over the airport itself. Meredith was directly below me in this freak winter thunderstorm, waiting for my safe landing ten minutes later. Nature's own free energy display was my greeting home.

All these demonstrations of humans and nature began to enlighten me with a broader truth about our reality. They set the stage for as many of us as possible to become empowered as Teams of Light to bring in the new paradigm. Then we may salvage our precious Earth from human desecration.

"For centuries the prophecies had looked to a change in consciousness as the unmistakable sign of a new age. This 'new mind' would set in motion the cultural forces that might bring about a true renaissance, an outer world that reflects inner harmony."

"Spaceship Earth, Gaia, a New Atlantis, planetary tribe, partnership, community, holism...these are not just ideas whose time is come. These are visions that transcend time. The concepts that are emerging now have the potential to shape a planetary civilization."

–Marilyn Ferguson, author of The Aquarian Conspiracy

10

Empowerment

Home, Ashland, August 27, 1994

Necessity is the mother of invention. Scores of brilliant scientists, inventors, engineers, physicians and others have recently become enlightened as individuals about the significance of the Consciousness Revolution and their part in it. And yet we have not yet seen these concepts expressed in the mainstream culture, which holds most of the money and power. My extensive travels and reflection have revealed to me a profound crisis of the spirit worldwide and a shocking decline of the condition of our environment.

I now hear the gentle sounds of nature's breezes in the trees being punctuated by the incredibly loud tearing of wood by a neighbor's power saw that had awakened us this morning. I feel timidly helpless about all the seeming attacks on our senses in what was going to be a retreat. I sit with parched lips and look at yellow brush tinderboxes on the land from the greatest drought and fire danger in Southern Oregon's history. I am experiencing a taste of apocalypse into an unhappy alternative future.

I think of the classic 1960s science fiction movie *Soylent Green* with Charlton Heston, in which virtually all life forms in the world except people have disappeared and the Earth is warming from a growing greenhouse effect. (As I mention in the Epilogue, recent evidence from El Nino conditions and reduced plankton, extreme weather patterns and the spreading of airborne diseases, have suggested to leading mainstream climatologists that global warming from burning fossil fuels has already begun in the Pacific region.) Set in the year 2022, we see a highly controlled patriarchal society in which Third World squalor and overpopulation have multiplied themselves and reached everywhere. The only remaining food that had any taste at all, soylent green, was rationed out once a week. But the substance, which was advertised to be a form of algae, turned out to be human remains processed in secret. Many people volunteered for their own deaths in exchange for a twenty–minute experience in a screening room showing the sights and sounds of a nature now extinct. Could this be our future?

Measuring versus Solving Global Warming

In the "real" world of 1995, the Scripps Institute has planned to install for the Department of Defense a set of suboceanic loud speakers and receivers that would measure changes in ocean temperature to measure the extent of global warming. Environmental groups have pointed out that these speakers, which would blare out a bunch of electronic buzzes several times a day, every day, off the coasts of Big Sur and Kauai, would almost certainly kill several thousand unlucky whales who happen to venture near. At this writing, the project is on hold.

The irony is, we seem to be intent on intruding on the environment in order to measure, not remove, another effect on the environment. This crude treatment of Gaia and some of her beloved creatures is analogous to doctors doing invasive diagnostic procedures on our bodies favored by some of our high–tech specialists. But the most supreme irony comes from the fact that for the cost of the Scripps project, we could do more than *measure* the problem of global warming. We could *solve* the problem of global warming by developing free energy. I am working with environmentalists to stop this misguided project.

Sooner or later, something will have to give. We will need to develop those social institutions which can facilitate the needed changes, first during a chaotic transitional period, and subsequently, when a new order can come out of the chaos. All this will no longer be politics as usual, and the job seems monumental. We need all the help we can get. Analogous to the founding of the United States over two hundred years ago, we must come together as Founding Mothers and Fathers at the table of *The Last Supper of Gaia* to form a new nation for all humanity and its cohabitants.

Meredith and I have both discovered that, in moving over to the right side of the table of Gaia, the deep–seated personal emotions of grief have still not vanished. Both grieving and transforming represent an ongoing process of stepping into the big changes ahead as we continue to observe the decline of our inner and outer environments. The neighborly buzzers, power saws, news reports of the evacuations and destructions from fire, all remind us of the real world where destruction seems to escalate.

Alas, I have found I cannot celebrate the new paradigm yet, simply through the process of accepting, creating, and becoming enlightened. The results are still mostly ignored by the larger society. This has been a dis–appointing but a necessary stage of my own transformation.

Looking at your example of a paradigm shift, how have you experienced your own phases of acceptance, enlightenment and creativity?

According to the research of Elisabeth Kubler–Ross and my own experiences, acceptance seems to include a residual sadness about what is actually happening, a quiet facing of the truth, and a look at the void without yet recognizing what might be revealed from within it. The creative and enlightening phases can begin to uncover that miracle but can also encompass the hard, lonely and unremunerative work of communicating products or concepts which are not yet ready for the marketplace or are culturally acceptable—whether they be inventions, books, paintings, speeches or whatever. Managing a business and supporting a family can be a full–time job that distracts many of us from our creative works, even if we have accepted the changes and entered the new paradigm. These are the times we most feel like "strangers in a strange land," like the alien visitors who might wonder about our denials and our ecocide.

Only by resonating with a mutually supporting network of co–creators, backed by the principles of a sacred science, can we truly come into the empowerment we will need to form a new society. I have found for myself that a resonance with others' enlightenment can also often lead to giving our power away to those who become gurus or experienced politicians in a traditional hierarchical male mode. That model will not suffice. On the other hand, we can learn a lot about what to do from ongoing scientific experiments on the creative influence of a collective consciousness in the material world. They can assist us in formulating those social structures that will bring in a new sense of empowerment for all of us. To back up these arguments, we now look at a remarkable set of relevant results which come from experiments carried out at one of my alma maters—Princeton University.

The Princeton Experiments on Bonded Couples

In *The Second Coming of Science* I devoted a chapter to the pioneering work of Robert Jahn, Brenda Dunne and their colleagues at the Princeton Engineering Anomalies Research Laboratories. These highly credentialled scientists have produced unquestionable evidence for the existence of operator influences on the results of random event generators (REGs) through the actions of their thoughts. These statistically significant and often repeatable experiments show that most all of us have the ability to use psychokinesis in our own unique ways.

In a more recent study on a variation of this theme, psychologist Brenda Dunne has looked at correlations between REG results and the intentions of fifteen co-operator pairs.

"The composite performance," she writes in her abstract," of eight operator pairs of the same sex is opposite to intention, while that of seven opposite-sex pairs conforms significantly to intention, with an average effect size 3.7 times larger than that of the single operator data. Of the opposite-sex pairs, four 'bonded' couples achieve average effects more than twice the size of those of three unbonded pairs, and nearly six times those of the single operators."

While Dunne cautioned against premature extrapolation of the information to all situations and operators, the gist of the results is obvious: coming together in bonded groups, we may be able to amplify the effects of our consciousness and accelerate the onset of a societal consciousness revolution. These seminal experiments suggest we can find environments in which bonded couples and other yet-to-be-determined confluences of individuals and pairs can create powerful changes in the material world.

These results are also consistent with the Klingbiels' and others' studies of the power of prayer and of meditation by groups to produce miraculous changes, whether they be a healing of an individual, a community or the entire planet. Sai Baba's abilities may draw, in part, on the support of his followers. I find I can demonstrate spoon–bending more effectively if I draw on the support of my audience.

One possible interpretation is that a resonant or morphogenetic field has formed around the intention of bonded couples to produce mind–over–matter effects about six times greater than they could achieve as individuals. Could you imagine how much more powerfully we could help change the world by using groups who are bonded and by creating the proper environment for them to act?

These kinds of experiments are a powerful tool that could give us a missing link to our empowerment. This is the kind of science we need to do a lot more of! The results also suggest that the relationships among those of us who want to be creative and enlightened about a new paradigm are important. Groups will need to work together; no one or two or a handful or a guru–led enclave will be able to fulfill the miracle in the void.

James Lovelock's *Gaia* and Peter Russell's *Global Brain* are concepts I described in *Exploring Inner and Outer Space.* Both works present scientific evidence that we humans could create a unified consciousness of great potential. Is it possible that, through our collective efforts, we could empower the global paradigm shift? These hypotheses represent an ultimate consequence of Dunne's experiments.

These metaphysical hypotheses also suggest that the entire planet could soon go through a major phase change, as if we were cells of a living Gaia. This superorganism we call Earth and our own personal selves are now in a chaotic, turbulent and dissipative state on the verge of collapse. According to the new theories of a sacred science, such an unstable state may then qualify Earth to emerge into a magnificent new order, as if from a caterpillar to a butterfly. All that is needed is for the system to be stimulated by some form of coherent energy.

The ideas of Prigogine, Sheldrake and others support this notion that we can trigger nonlinear changes by combining and collaborating on our intentions as a conscious effort to transform the world.

Some Speculations and Hypotheses

The scientific evidence is obvious. Coming together in teams, we can do wonders. In today's seeming dark age, wonders seem to be needed to lift the Earth and its cohabitants out of chaos and into a transcendent state. But parts or all of our superorganism might die in the process. In the coming years we will need to sharpen our metaphysical hypotheses into new models of reality. In and of themselves, neither free energy gizmos nor new science researched in a governmental or privately funded laboratory will suffice. We will need to develop a moral and spiritual fibre, to connect with a higher will, to help redirect our consciousness toward a healed collective consciousness.

The new agenda cannot be forced; it must be sourced from a higher place. Brenda Dunne's work points to higher success of bonded couples in having consciousness alter the material world and the Spindrift work suggests that the surrender to the higher will is a more potent force than using the wills of our egos. In other words, the most effective transformational tool will be teams of bonded people who surrender themselves to a Divine Will, to resonate with the intentions of the universe.

We also appear to be getting some help from the outside. Various allegedly channeled works such as Barbara Marciniak's *The Bringers of the Dawn* and *Earth* suggest that the oft-reported Pleiadians and other extraterrestrial groups are acknowledging and assisting us in our grieving. These apparently enlightened aliens have emphasized how important it is for us to feel, to surrender, to open our hearts, to form groups, and to speak.

"You are going to broadcast a frequency," said the Pleiadians channeled by Marciniak, "and that sound is going to travel. It is going to become a desperate aching and longing for the return to harmonics within the human race—a return to the power of the group mind and simultaneous empowerment of the individual."

Even though scientific details of such channeled information remain to be verified, the essence seems to ring true, and the scientific evidence continually points in that direction, especially in cases where we can observe prophecies being fulfilled, such as those of Nostradamus, Edgar Cayce, Gordon Michael Scallion, the Mayans and the Hopis.

It therefore appears that outside intervention in these "end times" may be occurring but on a limited basis, because of a policy of noninterference with our karma. This metaphysical hypothesis states that we create our own realities and that we ourselves are responsible for balancing our actions. These cause-and-effect relationships transcend our currently accepted science of materialistic determinism. Perhaps the concepts of karma and natural law, of which most humans seem to be unaware, are keeping alien cultures away from most of us while we exercise our free will to create our own realities.

As a result we can either have heaven on Earth or hell on Earth. The grey aliens and their cohorts seem to be using some of us for their own ends, but they have proven to be much less of a threat to life than what we are doing to ourselves and to other living beings. But the contact experience can still be an emotional whirlwind for those who have been abducted. It also provides abductees an opportunity to grieve the old paradigm and to begin to accept the new.

Other more positive contact experiences have enlightened many individuals to lay the foundation for the necessary changes. Perhaps the aliens themselves are demonstrating various alternative futures for us. Some investigators have suggested that our extraterrestrial and/or interdimensional visitors are not only our genetic engineers; they may be aspects of our own futures! Given the wide range of contact experiences reported by tens of thousands of witnesses, it is timely for us to look at some radical and relevant metaphysical hypotheses to be explored by new scientists.

The millennial prophecies and channeled works consistently point to the dramatic times we are now in. They acknowledge the existence of natural law and of the individual and collective soul. They talk about the schism between a degraded, dying and decadent Earth and a transcendent Earth. This apparent contradiction in terms suggests to me a more radical metaphysical hypothesis—that, at least in some respects, our world is bifurcating, or dividing into two or perhaps more forms, to accommodate the varying karmas of a rapidly changing culture. The old system could die and decay as the new system takes its place, like a snake shedding its skin or a butterfly discarding its cocoon.

Systems theory, chaos theory and the theory of evolution include so many instances in which these things actually happen, it would be no surprise to me to see Gaia do the same. What would the new Gaia look like? I do not know. What I do know is that unprecedented Earth changes, responding to escalating cancerous human activity, are beginning to overwhelm us. Extreme weather and uncanny earthquake and volcanic activity are signs that things are brewing. None of us really knows what will happen physically and what a planetary splitting might look like. All this seems to be a mystery only few of our enlightened alien prophets seem to want to reveal on certain occasions. It's as if we are not ready yet to be inducted into a cosmic awareness club. Meanwhile the waiting period seems lonely.

The most basic metaphysical hypothesis for a sacred science is that we each create our own reality, and to a degree, we all create our own collective realities. A hypothetical higher individual self and collective higher selves are the cosmic choreographers that orchestrate these opportunities from which we may learn and grow. Getting our consciousness into the driver's seat by extrapolating from the experiments of Dunne and the Klingbiels suggests we may be ready to act on these hypotheses. A compassionate superiority to a material world could give us a superior material world. The keys appear to be our bondedness in intention which aligns or resonates with a higher, more universal will. This *is* the Consciousness Revolution, which could also be considered the manifestation of what we call God.

Learning from the Founding Fathers

Many historians have examined those factors which empowered our forefathers to create the United States of America. When looked at from the point of view of whole systems theory and natural law (their own contemporary version of a sacred science), we can perceive some of their distinct attributes that may help empower us in our new historical mandate, free of the cacophony of contemporary politics.

About ten years ago I collaborated with legal scholar Joanne Gabrynowicz on taking a look at what we could learn from these extraordinary characters in history, in order to assist us in finding equitable ways of defining our own futures in inner and outer space. Gabrynowicz identified four distinguishing features about the founding fathers, which her research showed gave them the power to form a new nation. First they consciously exercised their free choice to remove themselves from the tyranny of the English king and to claim their inalienable civil rights. Secondly, they were close to nature, and so were aware of the natural order whose basic principles of cooperation and mutual governance they could apply to the laws of the land. Thirdly, they were spiritually motivated, allowing them to surrender to a higher will and to have the courage of their vision when human ego failed. And fourth, they were aware of the intimate relationship between the part and the greater whole, and so were able to form what eventually became the world's most powerful government. They achieved this by inventing what was then a stable blend of representation and checks and balances.

But in recent decades, when greed and secrecy got into the picture, the powerful Governmental structure, which once consciously reflected the natural order, began to fall ill. First Lady Hillary Rodham Clinton has spoken about our national crisis of the spirit. What has once been perceived as a mandate for our "inalienable rights of life, liberty and the pursuit of happiness as self-evident truths" seems to have become a bureaucratic and polemic nightmare of special interests which have hidden those truths. Washington appears to have turned mad in its own adversarial stew and has little to offer as real positive visions.

Bringing Together Our New Founding Mothers and Fathers

The ecological mandate absolutely requires that we dust off the works of our forefathers and add a few new ingredients. These include feminine participation and the openness to the emerging truths of alternative health, free energy, alien contact and psychokinesis. Perhaps most important is setting the goal of restoring a sustainable Earth as soon as possible. The larger natural system to which we need to become reunited is Gaia herself.

Our new founding fathers and mothers will need to sit at the table of Gaia and let her speak. This means a total shift in our science, philosophy, economics and significant parts of religious belief systems. The grieving process is deep and intense. The new nation will be a new Gaia, with us humans acting as conscious cocreators with the planet herself.

Where human will and activity are involved, the coming changes will take awhile. We have seen what a great task it will be for us all to move beyond denial. Nevertheless, the Consciousness Revolution will need to occur much more quickly than it took for the architects of the American nation to unfold their vision, which consumed over 150 years. The inevitable paradigm shift, the mother of invention, gives us only a few years to decades to carry out most of the shift to a truly sustainable global future. That time could be shortened by a significant consciousness shift that might at first appear to be an economic, political or natural cataclysm. Grieving and transforming could still consume a few years to give people an opportunity to redirect their own priorities and aspirations.

I suggest that new founding parents gather by the thousands to millions through our high speed communications systems, networks and local groups to help form a new, interdependent Earth–nation. First we will need to formulate a transitional government that will give us as much freedom as possible to make the necessary shifts.

I propose that a taxpayer–supported transitional government do some of the following, as a "strawman" scenario for a future Government:

• Adopt as its main policy the support of all those who will be most affected by the shifts.

• Help at all levels to provide basic human needs such as food, clothing, shelter, energy, maintaining the peace and personal support and education during the times of change. This would be very important in the event of economic collapse.

• Establish some tentative priorities for the new paradigm, such as global sustainability, health and free energy.

• Dissolve most federal and state bureaucracies and eliminate all jobs that don't contribute to the new.

• Redefine what is meant by "job" and replace no-longer-needed positions with those of part-time public service that are consistent with new goals, and where everybody participates. Government spending on current infrastructure could be cut tenfold at the federal level, over the transitional period, maintaining only those programs which are essential to our continuing well-being, and creating new programs that would accelerate the new.

• Reveal secrets relating to alien contact, free energy, etc., without prosecuting, except for those cases where civil rights were violated.

• Pass an amendment which would prohibit Governmental secrecy and disinformation except in those cases where national security is truly threatened. Government needs to be held accountable to the truth, no matter how painful it might be.

• Establish training programs that would help with the transitions, e.g., economic conversion plans from weapons to ecological clean–up projects, etc.

• For those who have "lost" their jobs, have the transitional government pay them on a gradually decreasing wage scale over about three years so they have the opportunity to grieve, retrain and perform part–time public service.

• Begin to eliminate world hunger, population growth, crime and plagues by advancing bold new programs that would look at root causes rather than symptoms.

• Step up research on new science projects to free our energies to evolve and mature into the new global village.

• Invite other nations to participate in the process.

• Uphold and enforce the Bill of Rights with no violence.

• Delegate most decisions to the local level, except for modest efforts to develop free energy and new science R&D, disaster relief, and keeping essential services and supplies going.

• Disband after three to five years.

During this time, we would form a council of elders and advisors who are knowledgeable about the new paradigm. These individuals would be self–selected so everyone can participate. The group would self–organize and then begin to establish the principles of our new agenda. They would convene for those three to five years in which the transitional government is in power, in order to build a lasting governmental structure and to begin to set new priorities, based on consensus. This group would disband after a "permanent" government is elected to implement its policies. None of its members would qualify for a government position for at least five years, but could serve in an advisory capacity. No single human being would have a term of office exceeding five years.

I am sure we all can think of additional ideas about making as peaceful and supportive as possible the coming chaotic transitional period. This time we will be letting go of old, outmoded ways and be going through the challenging stages of grieving and transforming. But, as Marilyn Ferguson has so wisely suggested, we don't want to mistake the process of change for its end. After a few years of shifting, we could then be in a position to reach a new consensus guided by the very powerful concepts of having a sustainable future under natural law. As tools, we can bring in the enlightenments of whole systems theory, new science, consciousness, free energy, alternative healing and alien contact.

To help guide us into these new initiatives I propose we develop a sort of manifesto which contains self–evident truths including those which motivated our forefathers, and expanded to our current context. The first draft of such a manifesto might read like the comments in this chapter and be circulated as a petition to be signed by a majority of voters. Then, the United States can once again become an inspiring example to the world as a leading democracy which is aligned with idealistic human values.

Through giving the power back to the people, the power of truly integral government and free energy, we Americans could restore our status as an example to the world as the good folks, directing us all into a sustainable future and a Golden Age. Because we don't have a Hitler or aggressive Soviet society breathing down our throats, we have an unprecedented opportunity to lead the world into an awesome future. The only enemy holding us back is ourselves. I believe we are coming closer to the time when we can kick out the professional politicians, the special interests, the practitioners of an old science, and the bureaucrats. There is plenty to go around, so let's start anew!

American inventor Sparky Sweet demonstrates one of his conditioned magnets used for his solid state free energy research. A former aerospace engineer, Sweet and I had met in 1968 while I was in the astronaut program. Sparky recently passed over. He will always be remembered as a great pioneer in this field.

During a recent trip to New Zealand, I took this picture of Bruce DePalma (above), the expatriated American inventor of the N-machine.

Dr. DePalma's latest—and confiscated!—N-machine (above right) is designed to produce kilowatts of power, with a potential to scale up to megawatts. When the rotor (left) is rotated, a voltage appears between the two mercury-wetted contacts to the right. Output power is dissipated by a water-cooled resistor (foreground).

The beryllium-copper alloy rotor (bottom right) has magnets at the ends and is designed for speeds of up to 30,000 revolutions per minute.

Dr. Paramahamsa Tewari, chief engineer for India's largest nuclear power construction project, has been given time by the Indian Government to develop this N-machine, which reliably produces 300% output/input power, using alternating current. These demonstrations were the most convincing proof-of-concept I had witnessed, and happened just before my visit with our common friend Sai Baba, whose photo hangs on the wall (above photo).

Dr. Shiuji Inomata of the Japanese Government's Electrotechnical Laboratories demonstrates his N-machine, which shows excess output power. He is now working with Toshiba Corporation on a commercial model using superconducting magnets. Inomata first inspired me with the diagram showing the importance of consciousness—a factor neglected by most physicists.

Consciousness

New Science

New Science

Matter

Mainstream Science

Energy

Canadian free energy inventor John Hutchison (right) shows what can happen to an aluminum bar when subjected to energy coming from his unique configuration of Tesla Coils and other equipment.

During a recent visit to our home, Hutchison (above) touches Meredith's sword onto a meteorite with the sun glinting. This man "communicates" with his metal, once causing a metal angel to melt anomalously on top of our wood stove. What was left of the angel is in Meredith's painting *The Last Supper of Gaia.*

Bulgarian–Australian–American inventor Yull Brown has found a hydrogen–oxygen gas mixture which produces extraordinary welding possibilities, as well as a highly efficient fuel for transportation.

Brown's technology includes the ability to create implosions which seem to involve the transmutation of elements and the safe release of energy.

Moray King (inset), also a systems engineer, authored a book on the theoretical aspects of the zero–point (free) energy. His work is based on principles understandable to physicists.

For many years, systems engineer Tom Bearden (right) has provided much of the theoretical and experimental basis for the efficacy of free energy.

The 1994 and 1993 (inset, right) Institute for Energy retreats at the Stanley Hotel in Estes Park, Colorado. Actor Jim Carrey (front row, third from the left) was filming *Dumb and Dumber* during the 1994 event.

Oil refineries like this one (above) are hazardous to our health. Even Christmas tree farms (upper right) are not immune from the grid system. Wind and solar power can provide some relief, but are too diffuse, capital–intensive and intermittant to compete with free energy ultimately.

We seem to be like sheep under power lines allowing the "lapis pig" consciousness to dominate our lives. But Japan is taking initiatives such as this 1995 free energy symposium graciously hosted by Tetsuhiro Aso (below). A lapis pig muppet scowls at the proceeding.

In Hiroshima (upper right), I gaze up at “point zero” of explosion in 1945 of the first atomic bomb ever dropped on a population. Fifty years later, John Hutchison (lower right) demonstrates a more friendly technology: his “crystal converter” free energy device wich produces continuous electricity with no moving parts.

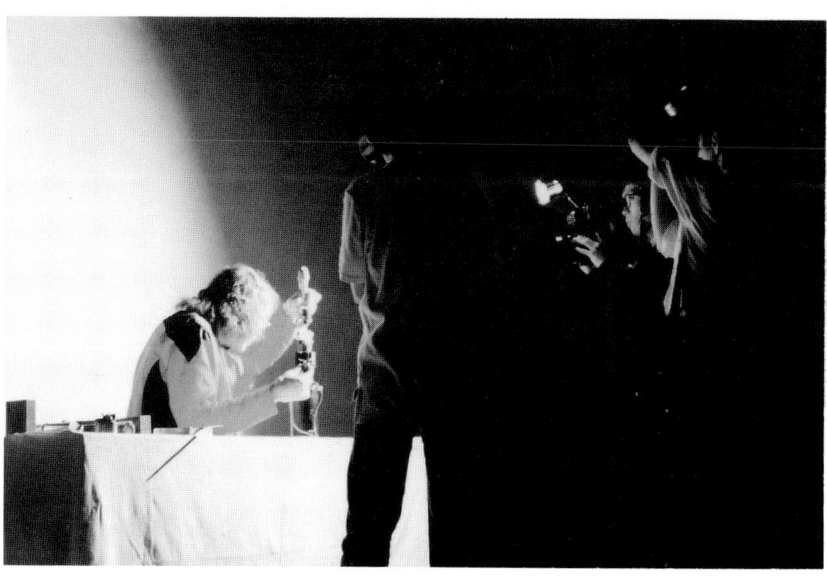

Meredith's painting *The Surrender of Man* symbolizes the need for man to lay down his sword of battle and conquest, to embrace the inner child.

Author Gary Zukav and I (inset) share a ceremony in the caldera of Wizard Island in Crater Lake to surrender to the wisdom of Gaia.

I took this photo of Bruce and John Klingbiel of Spindrift, just two weeks before their tragic deaths, following a long depression. Their outstanding research on psychokinesis and the power of prayer is dramatically confirmed by Sai Baba, (right) shown here producing holy ash (*vibuti*).

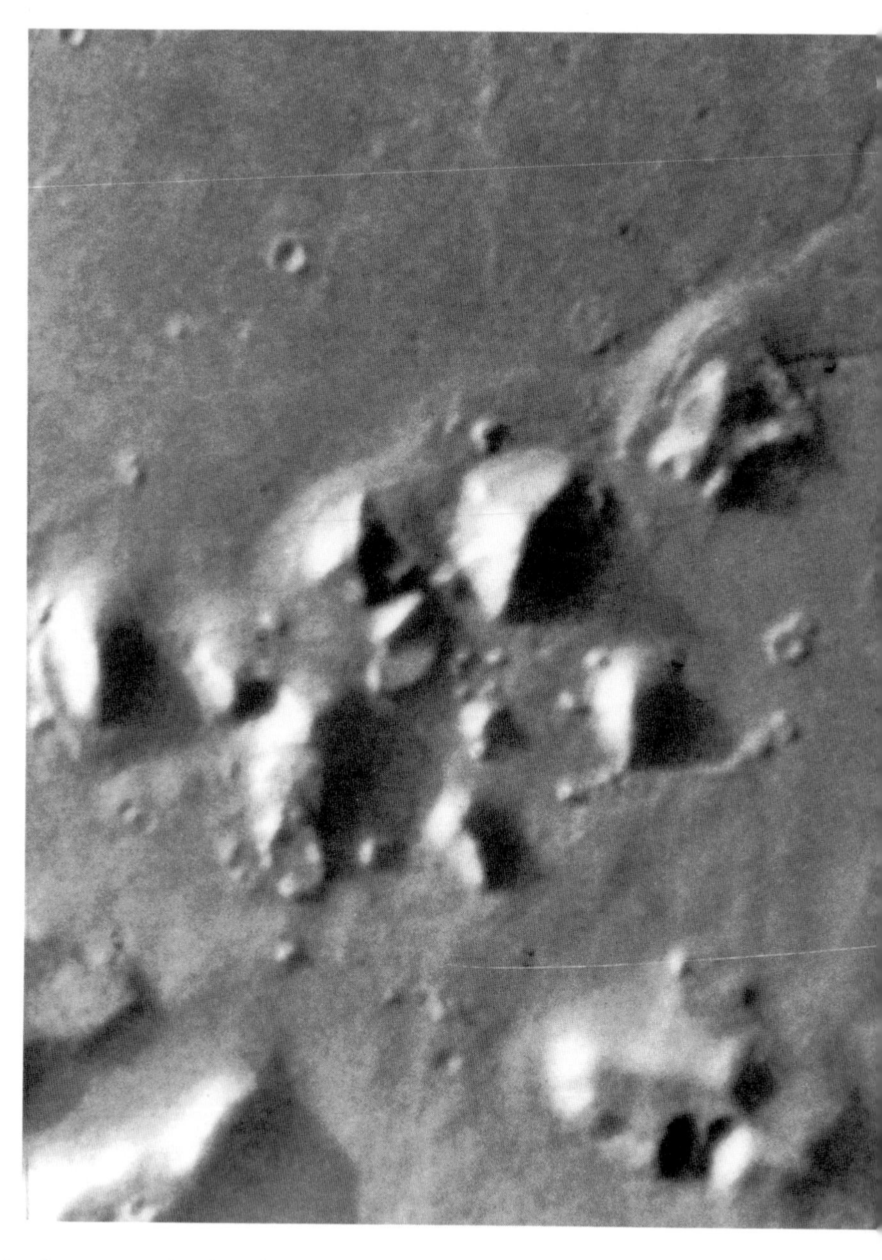

Stanley McDaniel (right), Professor Emeritus of Philosophy, recently authored a report on the U.S. Government's failure to properly investigate unusual features on Mars photographed in 1976 by the U.S. Viking Orbiter. For more than a decade, a number of us independent scientists suggested that these features, including the "face" (upper right) and "pyramids" (left) may be artifacts of a civilization. An excellent case study of two scientific paradigms clashing. Mars photo courtesy of Dr. Mark Carlotto.

I took this "New Eclipse" photograph showing two different forms of energy generation in rural Oregon: a windmill eclipsing smoke belching out of a power plant.

The Last Supper of Gaia by Meredith Miller. The disciples of grief sit on the left side while those of transformation are on the right side.

Each disciple represents a chapter in this book, as shown by line drawings in those chapters.

Central portion of *The Last Supper of Gaia* (left).

The artist poses in the autumn leaves of the California redwood groves, as she reflects on the plight and promise of Gaia.

The Resurrection of Gaia by Meredith Miller.

11

Transcendence

"...The transcendental experience is more real than the world of the senses."

–Deepak Chopra, in Ageless Body, Timeless Mind

"...We are on a journey toward authentic power, and... authentic empowerment is the goal of our evolutionary process and the purpose of our being. We are evolving from a species that pursues external power into a species that pursues authentic power. We are leaving behind exploration of the physical world as our sole means of evolution. This means of evolution, and the consciousness that results from an awareness that is limited to the five-sensory modality, are no longer adequate to what we must become."

–Gary Zukav, in The Seat of the Soul

I look over the words I have written, at the journeys I have experienced over the last two years of confronting the void. It has been an emotionally mixed blessing. On the one hand, I often feel great joy and satisfaction coming from gaining promising new knowledge and empowering ourselves into the new paradigm. On the other hand, the grief work often intervenes, and I'm wondering if feeling all those emotions might still be a part of the old paradigm itself, as a sort of healing crisis before the dawn. Or perhaps I am simply resonating with the pains of Gaia. I do not yet know.

Whatever this means, all this has caused me to contemplate, what precisely is the new paradigm?

In a word, it is consciousness. Scientists are clearly presenting us with powerful evidence, from many different directions, that our individual and collective consciousness can preempt and transcend the so-called laws of an objective, materialistic, time-trapped universe in which we are helpless, aging victims facing an inevitable and perhaps final death.

During this century, scores of laboratory experiments and other observations plainly show that we can influence the physical properties of anything "out there" in the material world by using our consciousness to interact with it. We ourselves seem to become inseparable from that force which may have created all of us in the first place.

To empower ourselves totally into this new paradigm is to have experiences and awarenesses that go beyond words. Free energy becomes more than cleverly designing machines. It *is* consciousness. The machines are merely an extension of that consciousness, devices to amplify what is already there and to make it work for us more efficiently.

In the new paradigm, the alien presence becomes more than tentative encounters with gray creatures from Zeta Reticuli or with space brothers and sisters from the Pleiades. It becomes a blending of consciousness between sentient beings.

In the new paradigm, physics becomes no longer fixated on particles that zing around in time and space. All these familiar variables fuse into a oneness we call consciousness, which is a more fundamental property of all creation underlying sacred science. The most recent observations and theories of physics make this principle even clearer.

The Most Recent Results in Physics

Perhaps the most dramatic contemporary development in physics has been a new recognition that mass may not exist at all! Physicists Bernard Haisch, Alfonso Rueda and Harold Puthoff have recently published a paper about these findings in the prestigious journal *Physical Review A* and a similarly high-brow publication, *The Sciences.* They conclude from well-researched evidence that "the concept of mass may be neither fundamental nor necessary in physics...The physical universe is made up of massless electric charges immersed in a vast, energetic all-pervasive electromagnetic field. It is the interaction of those charges with the electromagnetic field that creates the appearance of mass."

They go on to describe the nature of this field which physicists call the zero-point field (ZPF), which we have also seen to be the source of free energy. In his earlier presentations, Puthoff has frequently referred to the existence of this field from both experimental and theoretical perspectives.

"The amount of energy making up the ZPF is enormous," the three authors write. "That energy, in the conventional view, is simply forced into existence by the laws of quantum mechanics." They then go on to describe a simpler theory in which classical physics plus the ZPF may be all that is required to formulate a new paradigm of science. We can then scrap the complexities of quantum theory. "...The existence of a real ZPF is as fundamental as the existence of the universe itself," they write.

"The idea that space could be filled with a vast sea of energy does seem to contradict everyday experience. The answer to the question lies in the utter uniformity and isotropy (same in all directions) of the field. There is no way to sense something that is absolutely the same everywhere, outside and inside everything. To put the matter in everyday terms, if you lie perfectly still in a tub of water at body temperature, you cannot feel the heat of the water." Or it might be like trying to weigh a container of water somewhere beneath the surface of the ocean. What can you measure or detect from what?

The field becomes detectable only when an energy unit, or charge, is *accelerated* through space. When that happens, the charge experiences an electromagnetic force as a resistance to the acceleration. Instead of the Newtonian concept of inertia as a fundamental property of nature, the concept of mass dissolves into being a force exerted on an accelerating charge. Collectively, these forces comprise energetic fluctuations that can be detected experimentally, for example through the "Casimir Effect" which is observed in the laboratory. The forces are like a superposition over the apparent void of the ZPF.

What does all this mean? First, it implies we can tap the zero-point energy by interacting with the jitter of accelerating charges—for example, with a rotating Faraday disc, an N-machine or a specially conditioned magnet or crystal. Secondly, since mass is really only an electromagnetic force, then gravity itself must also be an electromagnetic force acting on it; otherwise things wouldn't fall. This means that antigravity propulsion becomes possible.

It also means that we can begin to truly unify the forces of nature without having to resort to such mathematically cumbersome formulations as string theory. "To reinterpret Einstein's equation $E=Mc^2$," the authors write, "we would say that mass is not equivalent to energy. Mass *is* energy."

Does this also mean that consciousness itself is also an electromagnetic force, in order for it to be able to interact with the ZPF? I believe that is so.

These heretical yet sound ideas, coming from three outstanding mainstream physicists and others, provide perhaps the most elegant theoretical framework for understanding free energy, gravity, and perhaps consciousness itself. Their published work was formally reviewed by no fewer than *five* peers (most papers require only two) and is certain to stimulate discussions among leading physicists. The time appears nigh for the radical revision of the foundations of physics, a movement which is also appearing in such unlikely places as the austere American Association for the Advancement of Science in recent West Coast meetings.

Science of the Vortex

The principle that matter is simply a form of energy has been independently proposed recently by the British scientist David Ash. Rather than accepting our contemporary cosmic zoo of fundamental particles and quarks as "things" which are the building blocks of matter, Ash believes that the particles themselves are energy trapped in vortices.

His arguments are persuasive. Analogous to the work of Haisch *et al*, Ash's approach reconciles many of the experimental anomalies and paradoxes of quantum mechanics. In his model, energy is a wave which travels at the speed of light, in agreement with traditional physics. Matter, on the other hand, is energy moving at the speed of light in a spiral motion that represents a vortex. As the vortex axis moves, the energy could be visualized as a spherical ball of yarn that superficially resembles matter.

"Both Einstein's and Newton's theories for gravity are weak because they contain arbitrary assumptions," says Ash. "Einstein assumed that matter distorts space–time in order to explain gravity while Newton assumed that gravity was caused by something nonmaterial acting between bodies of matter. Einstein never gave a clear picture of what space–time was so that it could be distorted by matter and Newton never left a clue as to what the mysterious nonmaterial entity was that acted between bodies of matter to pull them together."

As a physics teacher I have often explained to students that we do not know *why* gravity exists; we only know how it behaves mathematically. And we have not yet been able to unify gravity with other forces, a question being addressed by theoretical physicists who still deny the actions of consciousness.

"The Vortex theory," Ash writes, "begins with a clear description of space, time and matter. Their connection is that they are all mere aspects of the vortex. Because the vortex is curved, space is curved. Because the vortex extends beyond our direct perception, bodies of matter can act on each other at a distance (due to energetic interactions at the periphery of the vortices). Because the vortex is intrinsically dynamic the interactions can occur without the mediations of any nonmaterial entities."

Ash supports the notion that many of our nineteenth century physicists were onto the keys to the vortex as an underlying model for consciousness. It was during the current century we went astray in our denial of nonmaterialistic aspects of reality. He also presents evidence for a physics of ascension, combining these principles with those coming from the metaphysical and esoteric traditions.

Implications

It is fascinating and timely that some of the classical ideas of physics of the late nineteenth century are being given a new look. Such concepts as the ether and vortex are being dusted off and being incorporated into our thinking in light of new observations and theories about the zero–point field and the measurements of the actions of consciousness with the material world. The famous Michaelson–Morley experiment of the 1930s which verified Einstein's Theory of Special Relativity seemed to have eliminated the existence of an ether, but it was a deceptive result, because the kind of ether being tested was a "static" one that would not have picked up on influences on objects which *accelerated*. It appears that the physicists, in their zeal to verify the immutable properties of relativity, had thrown the baby out with the bath water. They had assumed that, because no effect was found for constant velocity or motion, that therefore there was no ether, or no zero–point field. We now know this misinterpretation of reality to be a mistake.

The implications of this new paradigm in physics are enormous. No longer do we need to worry about the cosmic zoo of particles and quarks which, according to quantum theory, are the fundamental building blocks of nature. Because matter doesn't matter, we see these little critters instead as discrete bundles of energy interacting with the ZPF. We can remove the many paradoxes inherent in quantum physics, such as the supposition that matter and energy can somehow be converted, one into the other. Instead, we have conversions between different forms of energy.

One of the surprises of twentieth–century physics is that most of matter is void space, in fact over 99.99 percent. Would it be that much of a stretch for us to make that number 100 percent?

We are at that point in physics when we either move into a more elegant Copernican system, metaphorically represented by simple orbits of planets around the Sun, or remain indefinitely in a Ptolemaic system, which places the Earth in the center of everything, and where the planetary motions are tracked by an unnecessarily cumbersome set of epicycles which never seem to converge into a solution. The old contemporary paradigm misdirects our attention to materialism and the belief that free space is empty, but now we know these "facts" are wrong. Any lack of interaction with the ZPF becomes merely a limiting case of an inactive application of consciousness.

Acknowledging and studying the zero-point field interactions—which is the source for understanding the genesis of free energy, gravity and consciousness—is an action akin to placing the Sun in the center of the solar system. The epicycles (the analogue to materialism) begin to disappear. They vanished entirely when later refinements such as Kepler's and Newton's Laws were brought forward.

Likewise, the current discoveries inevitably lead to a revolution in the sciences which will spread quickly. Combining these new ideas along with some others now budding, is bringing us into the new paradigm, into refining our understanding of consciousness and the ZPF—perhaps this *is* the Consciousness Revolution.

What About Time and Space?

According to Einstein's well-proven theory of Special Relativity, the passage of time converges to zero when one approaches the speed of light. This phenomenon gives rise to the famous "clock paradox" in which hypothetical travelers on a round trip from Earth into space at nearly the speed of light could return being younger than their own children. Decades may have passed on Earth while perhaps only months were clocked on board the spacecraft.

Based on this principle, author-scientist Peter Russell presents the following thought exercise in his recent book *The White Hole in Time*. He states that we all know that it takes nine minutes for us to receive light from the Sun because of the measured finite speed of light (c=300 million kilometers per second). But Russell poses an intriguing and inevitable consequence of relativity when we change our perspective to that of light itself.

"As far as light is concerned," he writes, "the moment it left the Sun is the same moment it arrived at my eye. From its perspective there is no time interval. This coincides exactly with my experience. The realm of consciousness and the realm of light would seem to share the same experience of now."

In other words, if you are light, you are everywhere at once. Any meaning to the idea of both time and space disappears, because the sunlight floods every point in the universe at the same time (from its point of view). As light, we have entered a new domain of reality.

Could the collective light of the Sun and other energetic sources in our universe also be at every point in space at all times? Is this possibly related to the zero–point field? *Could it be, the field itself is that light which is everywhere at once?* Are we inhabiting two realms at once— the familiar space–time domain, and the less familiar one of consciousness and light? Or shall we call it a new dimension?

Russell then proposed that matter is simply a white hole of "crystallized light," perhaps formed during the time of the Big Bang, and akin to Ash's vortex of energy.

My own hunch about the new directions of physics will include a good look at the heretofore elusive zero–point field and its interaction with consciousness. As the two concepts begin to fuse, they could possibly become the breeding ground for new technologies that will solve our most pressing problems of the environmment.

Perhaps my startling observations in India, about which I reported in the opening chapter, were only the start of what will soon become commonplace. Maybe we can *all* become empowered to extract energy and matter from the void. These activities could bring us into a transcendent dimension which only awaits our acknowledgment and exploration. We may then be able to learn to resurrect ourselves from a mortal and finite existence. Then we will become more fully conscious of our place in a universe fully alive and connected with who we are. This is what the new paradigm is about—transcending our self-imposed limitations of materialism, determinism and reductionism.

What is Consciousness?

My best definition is, consciousness is our intention to create something new in the universe. If that intention is properly aligned with the intentions of the universe itself, we will be able to create "miracles" of physical action that transcend materialistic determinism.

It is my belief that, as we move into the new millennium, the universe will be providing us with more opportunities to align, to accelerate the pace of synchronicity, to acquire Zukav's authentic power, to experiment with consciousness and its relationship with the material world, and to heal. *We now need to apply these tools to creating a globally sustainable future.* The rest of the universe appears to be far wiser than we are. Therefore, it will be we humans who will need to change, to surrender, to trust, to grieve the past, and to transform into the future.

Being able to align in intention with the universe seems to include our ability to tune into a resonant frequency, much like turning a dial to receive a station clearly on a radio set. Reports coming from Nikola Tesla, T. Henry Moray, Sparky Sweet, John Hutchison and countless other inventors show we have been able to occasionally tune into those frequencies which could produce more electrical output than input.

Yet the repeatability of their experiments has often encountered difficulties which has drawn no end of criticism from the armchair skeptics. Sometimes their devices produce little or no output, and sometimes, for no particular reason, they blow up, lose weight, take off, or turn to ice! Such demonstrations seem to reveal a burst or deficit of free energy and/or antigravity. Many free energy machines are now inexplicably unpredictable, almost as if the intention or consciousness of those involved might play a factor. Too many laboratories have reported too many anomalies to deny this reality.

Often the inventors have reported to me a number of mysterious effects that transcend the expected behavior of mundane electronics. These anecdotes, while not satisfying traditional scientific approaches, further verify the strange qualities of free energy. The stories are too often denied or dismissed because of our conservative bias, and provide a good analogue to the UFO abductee research of endangered Harvard professor John Mack and others.

For example, Canadian inventor John Hutchison "communicates" with metal, whose properties he has been able to alter with his free energy devices. On a recent visit to our home, he once focused his energy onto the fire of a wood–burning stove. When Meredith looked over at metal statues displayed on top of the stove, suddenly she noticed a pool of melted metal surrounding the face and wings of an angel which had sat there unharmed through hundreds of fires. This particular fire was not a hot one because the stove doors were wide open. Meredith placed what was left into her painting *The Last Supper of Gaia.*

The Relationship between Free Energy and Consciousness

David Ash and Peter Hewitt of England have proposed a theory of consciousness that involves what they call DNA resonance. According to this model, the helix (or vortex) is an archetypal form to bring in the needed resonances, much like tuning in a station using a solenoid coil and/or crystal. This time, the resonances are with our own DNA molecules, our consciousness acting like a radio transceiver.

Would you be able to amplify these effects simply by tuning a free energy machine to the wavelength of your consciousness? When you've *really* tuned into your station, perhaps your consciousness will burst wide open and you can become a Sai Baba, Uri Geller, Ingo Swann or Thomaz Green Morton. Or you have an explosion or levitation in your laboratory. Or you could create Backster effects, bio-communicating your consciousness to actions at a distance.

These phenomena and hypotheses suggest we bring together the experimental work in two seemingly separate fields. Could we combine the studies of consciousness with those of free energy, using statistical methods to assess the human–machine interaction to seek ways of producing electrical power *and* to create psychokinetic results through the actions of our consciousness? What might these fields have in common?

The psychokinetic/consciousness research includes the work of Jahn, Dunne, Spindrift, Radin, Vogel, Backster, as well as that of leading practitioners of mind–body medicine, such as Chopra. The energy research includes the theoretical work of a Puthoff, Bearden, King and Aspden, as well as the stories, models and results of experiments carried out by DePalma, Marinov, Inomata, Tewari, Sweet and many others.

Perhaps the applied science we sometimes call psychotronics could be born anew from tinkering with machines that would facilitate and amplify psychokinetic and healing effects, while at the same time providing appropriately-tapped energy. Blending consciousness with free energy could give us results we really need to produce clean electricity, antigravity and the potential to purify and transform our world in ways limited only by our imaginations. We can truly create our own reality.

In my opinion, consciousness will play a major role in formulating our inevitable free energy paradigm. In the new paradigm, free energy and consciousness may become just as inseparable as current-day electronics devices that provide a stimulating and comfortable sensory environment (light, sound, etc.). These technologies will allow us to enhance our inner and outer environments, both within and beyond five-sensory perception.

But if we pursue free energy from an old paradigm point of view and we continue to live in isolation and suppression, we may continue to get sporadic results. If we adopt a hypothesis that free energy *is* consciousness, the devices may then become tools to fine-tune and augment the intentions of that consciousness.

I have pointed out in *The Second Coming of Science* that the Spindrift/Radin work suggests we *can* design electronic systems analogous to biofeedback devices that mathematically filter the data in such a way as to amplify psychokinetic effects. Early experiments suggest that the degree of nonrandomness could be as large as one part in five for many operators.

Perhaps free energy devices will not end up being just electricity-producing black boxes made in Japan, while we Americans get dumb and dumber. Maybe the electrical uses are just an interim vision for the future, to be preempted by mind-over-matter technologies in which pure energy is a byproduct, a God-given resource like clean air, water and earth. Maybe our consciousness will become an integral part of a new fire that is safe and supportive of humans and the Earth—a more holistic form of free energy.

My foremost dream is to establish a consciousness/free energy/healing research and development laboratory for the benefit of humankind. This facility would integrate and augment these technologies to assist us into the Consciousness Revolution, free of bureaucatic and academic self-interest. It would also assist us in grieving the old and transforming into the new.

The Consciousness Revolution Seen Anew

The fact that we are still here now despite ourselves suggests there is a collective consciousness more powerful than that of any individual tyrant and his following. We have so far been able to avoid nuclear war, and the Earth is still vibrant in places. Perhaps, in consciousness, the universe may not allow us to destroy ourselves. Maybe we don't need to come from fear so much about the free energy genie coming out of the bottle. I believe there is a great hope, in spite of our tendency toward war and aggression. The experiments of modern science have made it clear to me that in the realms of light and consciousness there is no time or space—everything just *is*. But in order for us to have the experience of our perceptions, God entrapped some of that light into what we call matter, through consciously and creatively interacting with the ZPF, the implicate order, the void.

We had thought free space to be so empty and without consequence. How wrong we were. Now we know of a miracle in the void, the possibility to create untold universes. In our mass culture, the only universe that we have acknowledged exists is one that gives us the illusions of finite matter, time, space and physical death.

Our existence here seems to have created for us a collective vortex, a consensus reality of materialistic illusion, one that appears to be destroying our corner of the universe—Gaia. The inner and outer pollution we experience is our karma, and it seems we are at a great crossroads. We could either purify and reverse that debt or perish in our own stew. Federal, ecological and personal deficits will need to be balanced one way or another.

Our True Future

Through the ages mystics, prophets, religious leaders, scientists, ecologists and now channelers of disincarnate, extraterrestrial and interdimensional beings, have warned us that if we don't change our ways Gaia will do it for us.

But we also appear to be getting help from angels and aliens, apparently without interfering too much with our process. Perhaps their collective consciousness and prayers are blending with ours to help prepare us for the new universe we seek to embrace. Our transcendent friends appear to be watching us closely and are perhaps ready to help us when our global energy–economy–social–spiritual crisis becomes too unlivable for Earth and her species.

Could outside help be a dispensation of grace? Perhaps so. Even though the evidence for this hypothesis is anecdotal and sometimes subjective, I experientially and intuitively feel we are being guided and protected. We seem to be now entering into fulfilling those millennial prophecies which provide us with new options.

The greatest dilemma for me conceptually is how we will be able to transcend what often feels like a downward, not upward, spiral of karma. I often hope that the grief and remorse for what we have done is all that we need to express for us to quickly balance our past actions. Perhaps grieving itself is the payment of our karmic debts. If so, we will not need any more wars, crime, Earth changes, and patterns of aggression and greed. The Earth will not have to bifurcate. Once we do know our truth, we can collectively forgive the past as a final stage of our grief and beginning stage of our resurrection.

Meanwhile, our collective free will seems to have created an accelerating collective pattern in resonance with the destruction of our environment. How are we to awaken ourselves to the miracles of our own consciousness and reverse our actions against ourselves and the Earth? We must act soon.

The miracle in the void is that we can all empower ourselves to create beautiful new worlds, magnificent and novel universes. When we begin to resonate with the majestic and ubiquitous reservoir of pre-energy and pre-matter in the zero-point field, we will all become healers, clairvoyants and magicians. We can at last have peace, harmony, love and joy. Science is telling us that clearly, based on irrefutable experimental, theoretical and personal evidence. I invite you to trust the process and to walk with me through the visible into the invisible.

Epilogue

Home, Ashland, April 18, 1995

In the waning moments before this book goes to press, I look at the interchangeable snow and drizzle falling on drooping and dying daffodils. They, along with Meredith and myself, have been physically and emotionally challenged by the unexpected spring deep freeze that has been with us for more than a month now. Almost daily, unseasonal winter blasts have given us the flu, and we have been snowed in much of the time. We have a strong feeling of endless cabin fever, chaps, coughs and rashes.

We have had enough of the cold weather and isolation. Almost as soon as we listed our house, it was sold to the first customer who saw it. We will soon move to Maui. The bizarre circumstances of the weather keeping us inside seemed to taunt us to complete our projects. There is little else we could do with respect to the outside world as the deadlines of our three-year Oregon retreat rapidly approach us. The grouchy gloominess of our Easter winter was perhaps nature's way of underlining our own eleventh hour commitment to the end of a poignant period of painting and writing, a karmic call to action just prior to leading entirely new lives in the islands.

Even my dreams at night are becoming more about completion rather than the usual ones about trying to manage complex logistics (more characteristic of my recent life of much travel and the overwhelming responsibility of managing many things). During our recent scouting trip to Maui, after some health and comfort problems, we at last felt truly blessed as we settled into a beautiful hotel room (our third room in two days). Soon we viewed a complete, vivid rainbow arching over a blue Pacific Ocean with coconut palm trees flapping in a gentle and fragrant breeze.

Earlier that morning I had dreamt about completing a huge suspension bridge that spanned the Pacific from the mainland to Maui. It was a great feeling. A few nights later, a Maui friend had dreamt of me recklessly jet–boating (!) my way just offshore along the fringes of a glacier across a choppy ocean. Meredith and I both saw these dreams as clear signs we were completing things and moving away from the glaciers of Oregon to Maui as our new home. Perhaps our ashes work (grief), what Carl Jung had called "encountering your shadow," in Ashland was coming to an end, but not without a final exam.

As I now gaze out the window, as we dance along the final fringes of our creative projects, the glaciers outside begin to recede in the intermittent drizzle on a cold, sodden ground. This moment is exquisite in its melancholy, its feelings of wintry, fluey isolation, and our resistance to completion as Meredith faces her own resistances to completing her new painting *The Resurrection of Gaia* for a major Earth Day art show in San Diego this weekend. Tomorrow is Paul Revere (Patriots') Day, celebrated in my native Boston where I ran in the famous marathon 37 years ago. This had been an important local holiday commemorating the famous horseback ride of the patriot Paul Revere warning the townspeople and Minutemen that "the redcoats are coming". On that day, the battle of Concord began the American Revolution against the British in 1775. Tomorrow, on the 220th anniversary of that event, Meredith and I plan to climb on our horses southward to announce the need for a resurrected Gaia, a resurrected us, a miracle in the void, to lift us from the long gloomy winter of an Industrial Age.

And this book project comes to an end. The depth to which I feel that this too is a completion comes from looking at the unfolding changes and from reflecting on eight years of full-time travel and research on new science topics. Dreams, events, and the environment are telling me that *Miracle in the Void* completes a trilogy of books. The message in all three is clear: science must be expanded now to embrace a far greater reality. The new science can lead to technologies that will heal and transform ourselves and our planet. It is time to wake the town and tell the people, but not about a coming fiery Armageddon. As never before, we now have an opportunity to make a leap into greater truth. It is time that our ignorance and complacence come to an end, as Joseph Campbell had stated.

The most recent news about the global environment and on our karmic journeys makes this Paul Revere feeling ever more emphatic, as we attempt to herald the sleeping citizenry of nonlinear events to come—not only as an intuitive prophecy but as a scientifically-based set of predictions.

Home, Ashland, April 19, 1995

It is snowing hard here now, and we decided to stay for another day so Meredith could do the finishing touches on her painting. We would go to San Diego tomorrow all in one day, an 850-mile journey.

It is ironic yesterday's diary entry was written on the eve of the Oklahoma City bombing attack, which shows how not to express anger, how patriotism can be grievously twisted into the dark corners of reality. One of my best friends is an emergency room physician in Oklahoma City. He described to me that, after attending in a professional manner to a person who had already died, he went home and watched the television coverage. As the impact of the event began to

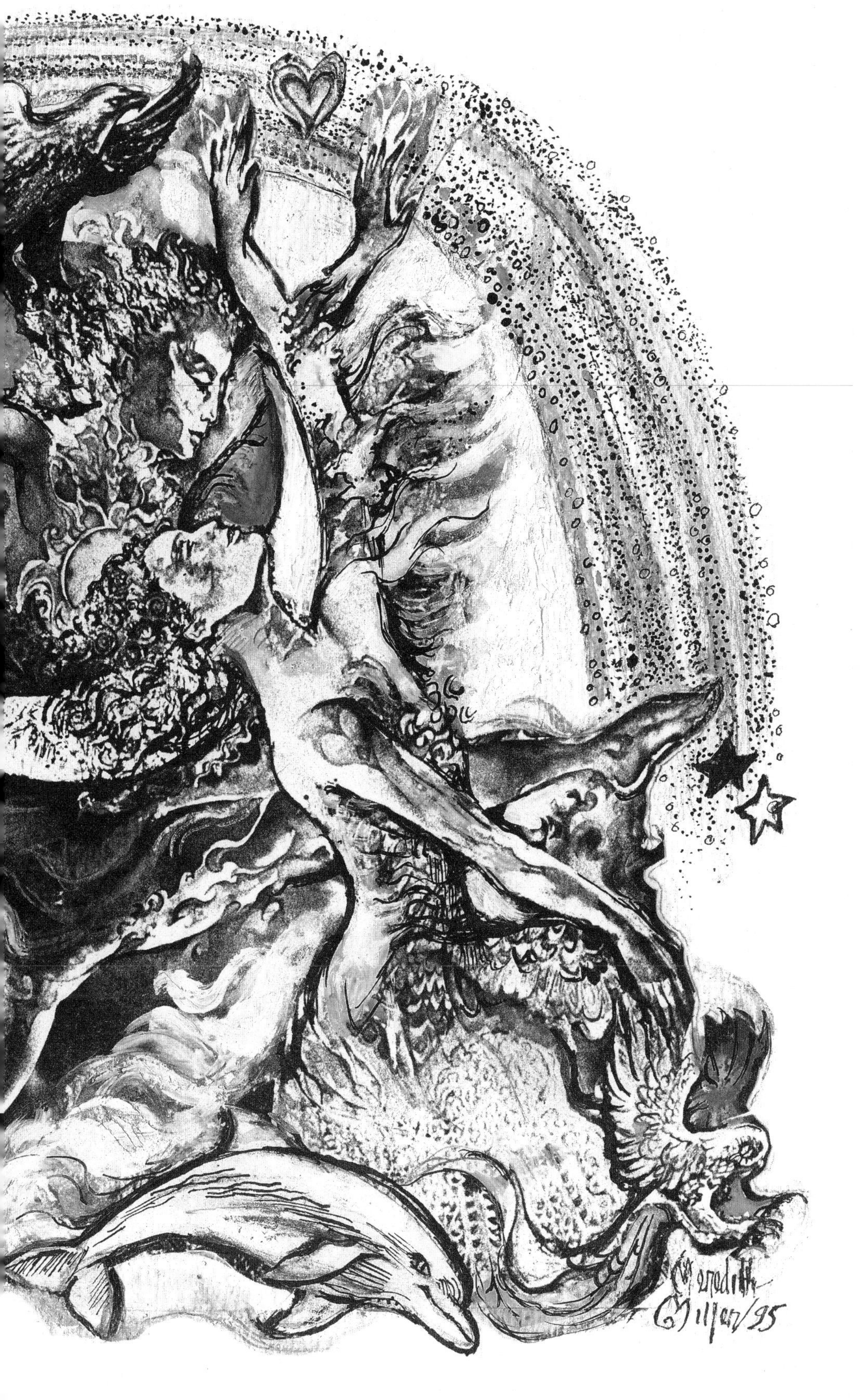
Meredith
Miller/95

dawn on him, he started sobbing, then cried deeply for hours. Here is the surrender of a man, an accelerated grieving, from which my friend now feels more liberated.

The Latest on Burning Greenhouse Gases, the El Nino and Global Plague Predictions

I was recently struck by two articles in the Washington Post (Outlook section, March 19, 1995) by Harvard public health physician Paul Epstein and journalist Ross Gelbspan, and by Jessica Mathews. Both pieces give evidence that our overburning of fossil fuels may have heated Earth's environment sufficiently to trigger plagues and other disasters.

Competent mainstream climatologists and disease control experts give us strong evidence that we humans are responsible for extreme weather conditions which have already made airborne disease more virulent and widespread than has been experienced during historic times—simply because we have relied on oil, coal and wood for our energy production. The unprecedented El Nino condition seems to stand in the middle of the causal chain between a manmade greenhouse effect and diseases related to climate change.

Unless the data are wrong, these circumstances require radical solutions such as free energy—ones that transcend the weak and incremental emission controls recently enacted at the Global Climate Treaty conference in Berlin. This gloomy news underscores the importance of moving immediately toward a free energy economy. We need an R & D effort now; it is clear.

Unresolved Personal Issues

We certainly do live in intense times. In recent days, Meredith and I could not have orchestrated more dramatic events that forced both of us to look at the most significant imbalances of the more–than–one century of relationship experience the two of us have collectively had on this planet. First, in our Oregon cabin isolation, we noticed we were beginning to repeat some of the old dysfunctional patterns of our own childhoods. In one particularly heated time between us, when we found ourselves repeating our parents' automatic interactions that blocked intimacy and understanding, an incredible phone call startled Meredith (I had just left the house at that time). The call came within minutes after the most intense of arguments we had had in our three years together.

This communication came from a former partner/spouse of mine. She stated she was now in therapy and still felt very hurt about a regrettable incident sixteen years ago, in which we had had a fight and I had slapped her in the face, resulting in bruises and a bloody nose. In this exquisitely timed call to Meredith, she described many of my basic shortcomings, some of them presumed to be unhealed. She wanted to go public about all this. I was obviously feeling very vulnerable after Meredith shared with me her lengthy phone conversation.

What would Meredith think of me with these stories? Fortunately, she perceived a higher truth and my former partner also began to see it: that we were all going through an emotional process of grieving, healing and forgiving ourselves and one another. This had been a one–time experience of rage, my own biggest skeleton in the closet, apparently coming up for me to look at in this intense moment. Even though my former relationship was very physical, volatile and emotional, the incident on that hot Labor Day night a long time ago showed me I could be violent without equitable cause. I was justifiably arrested, incarcerated and charged with a misdemeanor.

I had learned a valuable lesson that night on how not to express my anger. This reminder from the past could also have not come at a more poignant time, with a national focus on the O.J. Simpson trial and the shocking bombing of a federal building in Oklahoma City. Following the example of Russia, somehow a whole nation is viewing its own grief and transformation, through the expression of angry, aggressive, destructive male power in its last throes.

I had done years of therapy to forgive myself for the incident and I am grateful not to have come close to expressing physical violence ever again. But my former partner's hurt and humiliation for having been a victim of abuse was now just coming out and being shown to me—sixteen years later. Why now? Perhaps Meredith and I were coming closer to resolving, completing, and resurrecting our former pains, in resonance with current events on the news front. We were struggling with discomfort while karmic knots seemed to be sweeping through us, asking that these things be resolved soon.

This hypothesis seemed to be confirmed when Meredith's main former relationship imbalance also came forward. Her former partner turned out to live just two blocks from where we were staying in Maui. He had gotten wind of our presence there and called me in an angry tone about his perceptions of Meredith. Both Meredith and I were running interference for each other through some of our most major unresolved grief issues. It seemed that from out of nowhere we needed to address these questions.

Does all this mean that our karma is accelerating, reminding us of what is incomplete, and pushing us into a more rapid resolution of our personal affairs, so that we can feel more prepared to bring in the new? Could we now qualify ourselves to be among the growing ranks of Paul Reveres, taking the road-less-traveled of stretching into our pain during the dwindling moments of an old paradigm? Are we being strengthened by having the opportunity to experience our grief here and now so we can be better qualified to help lead us into a better world?

I hope so, and I hope forgiveness can come soon for us all, on both personal and global levels.

The Sun is now peeking through the clouds for perhaps the second time in two weeks. The glacier is melting. Meredith and I just spotted four deer on a rare grazing of our hillside. As medicine animals, deer represent the quality of gentleness overcoming our struggles. We are reminded that soft warm breezes soon await us.

Gaia is an awesome shaper of moods. We only need to believe in spring.

Chatsworth, California, April 26, 1995

We are now on the other side of the art show and entering into arrangements to have this book published. It is a hot, sunny, hazy day at Meredith's mother's house in a neighborhood which is evacuated for repairs from the 1994 Northridge Earthquake, whose epicenter is just two miles from here. (We had been here just the day before the earthquake, and so just missed it.)

The art show turned out to be a void in business terms—especially challenging because this was her first show in well over a year, and they are normally a major source of income. It had also been a most challenging one logistically. Meredith miraculously completed *The Resurrection of Gaia* after pulling two consecutive all-nighters, and then we did the fifteen-hour drive from Oregon to San Diego all in one day. We had left the house in the most intense snow-pour of the four winters we spent there, yet somehow we miraculously made it over Siskiyou Pass.

The show was in a beautiful Del Mar home. Meredith had just unveiled and interpreted her new painting, and I gave a global update to about 150 people gathered at the event. At that point the show had gone well and the feelings were electric. Meredith then led the crowd into another room, where she was going to interpret *The Last Supper of Gaia*. This particular presentation was to be recorded for use in this book, and was the climax of the show, prior to anticipated sales.

Just as Meredith was beginning to speak, one of our dearest friends, who is also an investor and participant in this project, was suddenly pushed by the crowd onto a large, fragile, unsecured marble table top beneath Meredith's important painting *The Surrender of Man*. The table tipped, crumbled into pieces, and dropped to the floor along with our friend and some valuable ceramics and crystal, which broke too. Fortunately he sustained only some cuts on his hand, but the incident took the electricity out of the event. Meredith couldn't speak. Once again Gaia couldn't speak. The owner was understandably upset. And another man had surrendered. People went home, and we felt the void as never before. Other accidents and near–accidents were happening to many of us involved in our emerging co–creation in art, science and literature. We were once again feeling a karmic knot moving through us—another dramatic opportunity to evolve.

It was at this point of void when Meredith and I both embraced a powerful lesson. In what might be traditionally construed as a business failure or stroke of financial bad luck, we decided instead to look at these events as an opportunity to "give from the void," in Meredith's words. We gave away some paintings, graphics and books to those who helped and to the owner of the house and broken treasures. We consciously decided to acknowledge the void and to keep a positive attitude.

And soon, new miracles began to fill the void. People came forward to assist us in our coming move. Friends exchanged services for paintings. And this book came together quickly. Relationships and dream-teams of empowerment became more important to us than mere money.

And so, on the shaken grounds of the San Fernando Valley, I reflect on the Patriots' Day Oklahoma City bombing, where rescuers are now searching the void spaces for possible miracles of life. I see the bombing as an escalation of the anger phase of our societal grief, another sign of the paradigm shift. Of course, we cannot tolerate this kind of violence any more than we can tolerate our rape of the Earth.

More than ever, I begin to recognize that the Consciousness Revolution awaits the surrender of man to a higher and more cosmic path. The Patriots' Days of Oklahoma City and Waco have been perverted by some to mean a duty to bear arms toward a violent revolution. But history teaches us that the real revolution is only heralded—and not consummated—by acts of violence.

With escalating dark there will be escalating light and life. This hope is more than a mere metaphysical hypothesis, speculation or metaphor. The growing light accessible to us *is* the miracle in the void. We only need to trust that process.

References and Readings

Albertson, Maury, and Margaret Shaw, *Proceedings of the International Symposium on New Energy, Volumes 1 and 2,* Room 203, Weber Building, Colorado State University, Ft. Collins, CO 80523 (1993 and 1994).

Ash, David A., *Science of the Vortex,* The Light University, 4 Western House, Station Road, Totnes, Devon TQ9 5LF, England (1993).

Aspden, Harold, *Proceedings of the International Symposium on New Energy* and *New Energy News* (1993 and 1994).

Bearden, Thomas E., *The Excalibur Briefing*, Tesla Book Company, Greenville Texas and Strawberry Hill Press, San Francisco (1998 and 1990); *AIDS as Biological Warfare,* Tesla Book Company, Greenville, Texas (1988). Other Bearden papers and books may be ordered from the Tesla Book Company or through Bearden at 2311 Big Cove Road, Huntsville, AL 35801. His latest works are published in the *Proceedings of the International Symposia on New Energy* (1993 and 1994).

Bly, Robert, *Iron John*, Addison–Wesley, Reading, MA (1990).

Bohm, David, *Wholeness and the Implicate Order,* Routledge & Kegan Paul, London (1980).

Chopra, Deepak, *Ageless Body, Timeless Mind,* Harmony Books, New York (1993).

Davidson, John, *The Secret of the Creative Vacuum,* C.W. Daniel Co., Ltd, 1 Church Path, Saffron, Walden CB10 1JP, England (1989).

Dunne, Brenda, "Co–Operator Experiments with an REG Device." (Technical Note PEAR 91005). Princeton Engineering Anomalies Research, Princeton University, School of Engineering/Applied Science (1991).

Eisen, Jonathan, ed., *Suppressed Inventions and other Discoveries*, Auckland Institute of Technology Press, New Zealand (1994).

Ferguson, Marilyn, *Brain and Mind Bulletin,* May and June, 1994.

Fox, Harold, *New Energy News,* P.O. Box 58639, Salt Lake City, UT 84158 (1994 and on).

Haisch, Bernhard, Alfonso Rueda and Harold E. Puthoff, "Beyond $E=Mc^2$," *The Sciences* (November/December 1994) and *Physical Review A* (February 1994).

Harman, Willis, "Toward a New Eco-nomics," *Noetic Sciences Review,* Autumn, 1994.

Holmes, Ernest, *The Science of Mind,* a classic book which is available at Religious Science Centers.

Inomata, Shiuji and Yoshiyuki Mita, "Design Considerations for Super Conducting Magnet N-machine JPI-II," *Proceedings of the International Symposium on New Energy* (1994).

Kelly, Don, *Space Energy Association Newsletter,* P.O. Box 11422, Clearwater, FL 34616 (1993 and on).

King, Moray, *Tapping the Zero Point Energy,* Paraclete Publishing, P.O. Box 859, Provo, UT 84603 (1989); also look for his contributions to the International Symposia on New Energy (1993 and 1994).

Kubler-Ross, Elisabeth, *On Death and Dying,* Collier Books, New York (1969).

Kuhn, Thomas, *The Structure of Scientific Revolutions*, Univ. of Chicago Press, 2nd Edition (1972).

Mack, John, *Abduction,* Random House, New York (1994).

Mallove, Eugene, *Fire from Ice,* John Wiley & Sons, New York (1991).

Manning, Jeane, *The Coming Energy Revolution,* Avery Books, New York (1995).

Marciniak, Barbara, *Bringers of the Dawn,* Bear & Co., Santa Fe, NM (1992); and *Earth,* Bear & Co. (1995).

McDaniel, Stanley, *Yoga Sayings*, Dep't of Philosophy, Sonoma State University, Rohnert Park, CA 94928 (1991).

McDaniel, Stanley, *The McDaniel Report,* North Atlantic Books, Berkeley, CA (1993).

Michrowski, Andrew, *Proceedings of the International Symposium on New Energy* (1994).

Morgan, Marlo, *Mutant Message Down Under*, MM Company (1991).

O'Leary, Brian, *Exploring Inner and Outer Space,* North Atlantic Books, Berkeley, CA (1989).

O'Leary, Brian, *The Second Coming of Science,* North Atlantic Books, Berkeley, CA (1993).

O'Leary, Brian, "Green Power: The Coming Free Energy Revolution and the Return to Eden," *International Symposium on New Energy*, ed. by Albertson (1994). This paper is also reprinted in Appendix II.

O'Leary, Brian, "The Suppression of Free Energy, UFOs and New Science," International Forum on New Science, ed. by Albertson, Fort Collins, CO (1994).

Puthoff, Harold, "Alternative Energy Sources: The Good News/Bad News and the 1–Watt Challenge," *Proceedings of the International Symposium on New Energy.*

Redfield, James, *The Celestine Prophecy,* Warner Books, New York (1993).

Russell, Peter, *The White Hole in Time,* Harper, San Francisco (1992).

Sheldrake, Rupert, "Seven Experiments That Could Change the World," *Noetic Sciences Review,* Autumn, 1994.

The Spindrift Papers, Spindrift, Inc., P.O. Box 3995, Salem, OR 97302 (1993).

Strieber, Whitley, *Communion*, Wm. Morrow, New York (1987); *Transformation*, Wm. Morrow (1988); *Majestic*, G. P. Putnam's Sons, New York (1989); and *Breakthrough* (in press, 1995).

Talbot, Michael, *The Holographic Universe,* HarperCollins, New York (1991).

Tewari, Paramahamsa, "Generation of AC and DC Power from Space Power Generators," *Proceedings of the International Symposium on New Energy* (1994).

Weaver, Dennis, "Ecolonomics: Fusion of Ecology and Economics," *Proceedings of the International Symposium on New Energy* (1994).

White, John, *The Meeting of Science and Spirit,* Paragon House, New York (1991).

Zimmerman, Michael, "Why Establishment Elites Resist the Very Idea of Non-Human Intelligence," Department of Philosophy, Tulane Univ., New Orleans, LA 70118 (1994).

Zukav, Gary, *The Seat of the Soul,* Simon & Schuster, New York(1989).

Resources

I recommend to anyone who wants to stay up to date on free energy developments, the following periodicals:

- *New Energy News* – a monthly newsletter for members of the Institute for New Energy. Memberships cost $35 per year. Edited ably by Hal Fox, Fusion Information Center, P.O. Box 58639, Salt Lake City, UT 84158, Phone (801) 583–6232.

- *Infinite Energy* – a bimonthly, slick magazine with a technical slant on cold fusion and other new energy technologies. Subscriptions are $29.95 per year U.S. and Canada, $49.95 other countries. Published and edited by Eugene Mallove, Cold Fusion Technology, P.O. Box 2816, Concord, NH 03302. Formerly called *Cold Fusion Magazine*. Phone (603) 224–5975.

- *Extraordinary Science* – the magazine of the International Tesla Society (ITS). Subscriptions cost $25 (U.S.), $30 (Canada/Mexico), $50 (others). Address is P.O. Box 5636, Colorado Springs, CO 80931, Phone (719) 475–0918, fax (719) 475–0582. The ITS also has a bookstore of free energy publications of historical interest.

- *Space Energy Journal* – a quarterly periodical on the latest, free energy hardware concepts. Subscriptions are $35 (U.S./Canada), $40 (others). Published by Jim Kettner and Don Kelly, Space Energy Association, P.O. Box 11422, Clearwater, FL 34616, phone and fax (813) 461–7119. Kelly also shares updates and correspondence with colleagues on political as well as technical information.

- *Electric Spacecraft Journal* – a quarterly on free energy related topics, thoughtfully edited by Charles Yost. Subscriptions are $24 (U.S.), $29 (Canada/Mexico), and $39 (overseas). Address is 73 Sunlight Drive, Leicester, NC 28748, Phone (704) 683–3511.

Also two magazines which report on free energy developments and are understandable to the lay person:

- *Explore! and Explore More!* – published bimonthly, these magazines provide up–to–date information about the latest in free energy and other new science developments. Subscriptions are $39/year US, $49 other countries. Published and edited by Chrystyne Jackson, *EXPLORE* Publishing, P.O. Box 1508, Mt. Vernon, WA 98272, Phone (206) 424–6034, fax (206) 424–6029.

- *NEXUS New Times* – popular bimonthly down–under/U.S. magazine exploring the frontiers of new energy, science and technology. Subscriptions are $25 per year US, $40 Europe. U.S. office, P.O. Box 177, Kempton, IL 60946, Phone (815) 253–6464, fax (815) 253–6300; main office, P.O. Box 30, Mapleton, Qld, Australia, Phone 61–74–42–9280, fax 61–74–42–9381.

Finally, for those of you technically inclined I recommend the *Proceedings of the International Symposia on New Energy* (1993 and 1994), edited by Maurice Albertson and Margaret Shaw, Room 203, Weber Building, Colorado State University, Fort Collins, CO 80523, Phone (303) 482–3731.

The best general journalistic piece on free energy developments is *The Coming Energy Revolution* by Jeane Manning, Avery Books, New York, 1995.

Glossary

abductions – A widespread phenomenon in which humans are taken away for a period of hours (usually) and examined by aliens. Several popular books by Budd Hopkins, John Mack, David Jacobs and others have reported details of this most curious UFO–related phenomenon.

alchemy – The ancient spiritual science of transmuting elements, ignored by most modern chemists and physicists.

alternative health – The growing branch of medicine which looks at healing, positive thinking and consciousness as a tool—beyond drugs and surgery.

animal mutilations – Reports of mysterious, precise removals of organs from animals, believed by some investigators to have been caused by aliens who are experimenting with genetic material.

anomalous – Unusual or unpredicted; excess.

anthropocentrism – The belief or philosophy that nature is raw material to be used by humans and that humans are at the top of the hierarchy of sentient beings.

anthropocentric patriarchal humanism – Philosopher Michael Zimmerman's phrase for our contemporary set of beliefs and assumptions which are blocking an accurate perception of our place on the Earth and in the universe.

antigravity – An application of free energy in which the effects of gravity can be counterbalanced, providing an elegant new propulsion system probably used by UFOs.

Area 51 – A top secret U.S. military facility where some investigators believe UFO technology is being tested.

ashes work – Author Robert Bly has suggested that men (particularly) go through a process of looking at their world view and shift it through grief and feelings. I've been doing this in Ashland for nearly four years.

authentic power – Author Gary Zukav's phrase for power sourced from within, aligned with the natural laws of the universe.

Big Bang – A theoretical explosion of matter and energy about 15 billion years ago, which may have created our current universe.

biocommunication – Telepathy.

black budget – Approximately $35 billion or more of U.S. tax money goes to these projects, which have no public accountability and involve various activities conducted in secret.

Brookings Institution Report – U.S. Government–funded report in 1961 of experts who recommended that evidence for artifacts discovered in our solar system or other signs of extraterrestrial intelligence be suppressed. Cited in *The McDaniel Report*, this may have provided a policy precedent for a Mars anomalies coverup.

Cartesian – A science which is totally based on materialism, determinism and reductionism, derived from Descartes.

Casimir effect – Physicist Harold Puthoff has cited this as experimental evidence for the existence of free energy.

chaos theory – A recent theory expounded by Nobel laureate physicist Ilya Prigogine, which presents circumstances under which order can come from chaos.

Chernobyl – Location of a nuclear power plant accident that occurred in the Ukraine, resulting in widespread radiation poisoning.

classical dynamics – The branch of physics which deals with the motion of bodies in free space subject to Newton's Laws of motion and gravity.

clock paradox – An aspect of Einstein's Special Relativity in which the time elapsed for those traveling near the speed of light is shorter than those left behind.

cold fusion – Chemists have discovered that certain palladium alloy electrodes dipped into a solution could produce what appear to be nuclear reactions and the release of excess energy. This is probably a form of free energy.

consciousness – Our ability to alter the material world through thoughts, feelings or prayer.

Consciousness Revolution – This is a revolution in thought occurring now (look at definition of consciousness).

Copernican Revolution – Nearly five centuries ago, the Polish astronomer Nicolaus Copernicus suggested the Earth was not at the center of the universe, causing a shift in our thinking about our place in the cosmos.

cosmic Watergate – A phrase often used for the UFO coverup.

crop circles – A mysterious phenomenon in which wheat stalks are flattened to form curious circles and other shapes, appearing mostly in Southern England.

darshan – Blessing.

Darwin – The scientist who first proposed the Theory of Evolution.

determinism – The philosophy of understanding scientific reality based entirely on the cause-and-effect relationships of matter and energy interacting in physical space-time (part of the old paradigm which is inhibiting us).

DIA – Defense Intelligence Agency.

disinformation – Lies, mistruths.

DNA resonance – A theory proposed by British scientists David Ash and Peter Hewitt that the DNA molecule can resonate with energy fields of consciousness.

DOD – Department of Defense.

DOE – Department of Energy.

dynamic ether – A turbulent substratum of matter–energy which represents the "jitter" of the zero–point field, or quantum fluctuations of the vacuum. Physicists eliminated the existence of a static (motionless) ether, but some erroneously extended that exclusion to a dynamic ether.

Dyson Sphere – Physicist Freeman Dyson speculated that some civilizations' thirst for energy would end up by their constructing spheres of matter entirely surrounding their central stars.

Einstein's Special Theory of Relativity – The theory which describes the distortion of time–space at velocities near the speed of light.

El Nino – Spanish for "child"; a warm spot in the South Pacific Ocean which occasionally moves eastward, causing climate changes on Earth.

electric jail – Our polluting grid system of powerlines. I often used this phrase as a young child. Perhaps I was in touch with a deeper wisdom about the times we were coming into.

electromagnetic pollution – Environmental and health effects induced by transmission lines and other electromagnetic equipment.

entropic universe – The old–paradigm opposite of chaos theory, in which many physicists believe the universe is becoming more disorderly and will eventually wind down.

epistemology – The branch of philosophy which addresses the question of how we know what we know in scientific investigation.

Faraday disc – In the early 1800s, the great English physicist Michael Faraday first discovered the unusual electromagnetic properties of rotating magnetic disks, which appear to produce free energy.

ferromagnetism – Phenomena which involve the use of iron magnets.

fossil fuels – Oil, coal and natural gas produced by the gradual decay, over hundreds of millions of years, of the remains of living matter. In this century alone, most fossil fuels have been burned up for electricity and transportation.

free energy – The zero–point energy which exists in the vacuum of space. Sometimes called new energy, space energy, scalar energy, vacuum energy, Tesla energy, or quantum energy (not solar energy). While it has been unacknowledged in mainstream circles, there is plenty of experimental and theoretical evidence that free energy exists and is tappable. This energy is extraordinarily abundant and could be used for clean and cheap electricity. The technology is currently in the research phase of the R&D cycle. Devices include cold fusion, rotating magnets, solid state systems, and hybrid systems.

fundamental particles – electrons, protons, neutrons and other small units of matter/energy.

Gaia – Greek for "Earth Mother," the goddess of our planet.

Gaian mutation – The possibility that the Earth will radically change, in order to adapt to recent stresses imposed by humans.

Galactic Club – Those civilizations which are hypothesized to be in contact with one another throughout the galaxy and perhaps beyond.

Gestalt – A hidden pattern which becomes suddenly recognizable, based on concepts of psychology.

global ecocide – Human–caused destruction of the environment.

gray aliens– The most common aliens reported by abductees, as sketched by Meredith Miller opposite Chapter 1.

greenhouse effect – The trapping of solar heat in the Earth's atmosphere attributable, at least in part, to carbon dioxide emissions from burning fossil fuels and wood. Climatologists believe this creates global warming and El Nino patterns and weather extremes that may invite the widespread breeding and spread of airborne disease.

grid system – The system of transmission lines running from power plants to end–users.

ground traces – Alterations of the ground from UFO landings or other interactions, which show unusual physical and chemical effects.

hot fusion – For decades physicists have tried to simulate the Sun's interior nuclear reactions by attempting to squeeze hot gases within a huge donut-shaped container. In spite of billions of dollars spent, we haven't yet achieved "breakeven" for energy output. Unlike cold fusion and free energy, a hot fusion reactor still requires large central power plants and a grid system.

hundredth monkey – Popular story of anthropologists' observations that the consciousness and behavior of monkeys of a given species throughout the world suddenly changed when one hundred of its members on a particular island learned a new habit.

hypersphere – A four-dimensional sphere, represented in three dimensions by a donut with a tiny hole in the center.

Information Age – The recent proliferation of information through the use of computer networks, broadcasts, etc.

Institute for New Energy (INE) – An institute recently spun off from the IANS that is the world's principal professional organization fostering the development of free energy.

interactive resonances – The process of consciousness interacting with physical phenomena.

International Association of New Science (IANS) – An organization I co-founded which holds annual meetings encouraging new scientists to present their results in a friendly environment with peers.

Iran-Contra – An illegal, secret U.S. Government deal involving Oliver North and the Reagan Administration.

karma – A natural law of cause-and-effect proposed in some Eastern religions to preserve balance in the universe.

Kepler's Laws – A fairly precise representation of elliptical planetary motions around the sun, as a refinement of the Copernican system.

lapis pig – My symbol for exploitive capitalism, the status quo, and the financial and scientific vested interests—with laughter!

Mars anomalies - Principally the "face" and "pyramids" in the Cydonia region of Mars photographed by the 1976 Viking Orbiter. The scientific evidence is strong these features are artificial structures constructed by a civilization.

megabytes - Computer information in the millions of units of information.

megaparadigm - A very large paradigm containing a subset of smaller paradigms; an evolutionary modality.

metaphysics - The branch of natural philosophy which explores the nature of reality; literally a greater, more all-inclusive physics.

McDaniel Report - Philosophy professor Stanley McDaniel wrote this report citing the failure of the U.S. Government to come to grips with investigating the Mars anomalies; a coverup is suspected.

military-industrial complex - First warned about by former President Eisenhower, this is our legacy of World War II and the Cold War that has created an infrastructure which is now controlling information and weapons.

millennial paradigm - The new paradigm of the next one thousand years.

miracle in the void - The principle that the vacuum of space is full of potential energy; that nothing is something.

morphic fields - Cell biologist Rupert Sheldrake has identified these energy fields of consciousness (morphogenetic) which tend to reinforce existing or new habits, transcending the materialistic deterministic assumptions of Western science.

N-machine - A rotating, magnetic disk free energy generator invented by Dr. Bruce DePalma.

Naessens, Gilbert - A Canadian inventor of a breakthrough microscope which has been suppressed, as reported in books and articles by Christopher Bird.

NASA – National Aeronautics and Space Administration.

new energy – Any newly discovered or acknowledged energy source.

new science – The science which addresses subjects considered "outside the box" by mainstream science, such as psychokinesis, alternative healing, UFOs, etc.

Newton's Laws – Laws of motion and gravitation which apply to objects travelling at speeds well below that of light.

over–unity – A phrase used by researchers for achieving a free energy threshold in which electrical output exceeds input, as measured by traditional means.

paradigm – World view; a consensus reality which dominates long historical periods.

parapsychology– The branch of psychology that deals with psychic phenomena and consciousness.

particle accelerators – Apparatus used by physicists to understand the nature of fundamental particles which comprise matter.

Perennial Philosophy – A phrase coined by the late British author–philosopher Aldous Huxley, referring to those qualities shared by the major religions of the world (one God, higher consciousness, the power of love, etc.).

perpetual motion – One result of tapping free energy is the ability of machines to run indefinitely. This phrase is often used by skeptics and naysayers to debunk free energy, because they erroneously feel this is a violation of the law of energy conservation.

petrochemical – Oil–based chemicals.

physics of ascension – The scientific basis for moving through our own evolution to higher states of being (esoteric).

Pleiades – A star cluster from which some channelers have reported a certain benevolent class of aliens are communicating with us.

polygraph – Lie detector.

precognition – Predicting future events.

pseudoscience – Pretending to be science.

psi – Short for psychic phenomena.

psychokinesis – Mind–over–matter; the ability to alter the physical universe through consciousness.

psychotronics – The science of devising machines which interact with human consciousness.

Ptolemaic system – The (erroneous, cumbersome) pre–Copernican idea that the Earth is the center of the universe, involving complex math–ematical constructs such as epicycles.

quantum theory or mechanics – The prevailing theory of physics underlying the behavior of matter at the smallest scales (atomic, fundamental particles, quarks, etc.).

quarks – The smallest understood division of matter being probed by physicists.

random event generators – Electronic devices which generate binary numbers (ones or zeros) randomly. They are now commonly used in modern experiments with human operators on psychokinesis.

reductionism – The philosophy of understanding scientific reality in terms of reducing matter to its smaller physical component parts (old paradigm).

Rife, Royal – Suppressed inventor of a microscope reported in the book *Suppressed Inventions*.

rotating magnet generators – Free energy devices which employ rotating disks containing magnets. Inventors include Bruce DePalma, Paramahamsa Tewari and Shiuji Inomata.

sacred science – The science which looks at ways in which we are interconnected (rather than dissected), and involving the phenomena of consciousness.

scalar energy – Free energy explained in terms of "scalar" waves (having no particular direction but very large magnitudes as opposed to "vectors").

Secrecy Act – This "Gag Rule" gives the U .S. Department of Defense the authority to confiscate free energy devices they feel have weapons application.

SETI – The Search for Extraterrestrial Intelligence program, focused mostly on radio telescopic searches for signals coming from hypothetical, intelligent civilizations within other star/planetary systems. Congress has cancelled funding for the program, and no positive results have yet been obtained.

solid state – Electromagnetic components with no moving parts which can be made into a free energy device (e.g., Sparky Sweet's and John Hutchison's devices).

Soylent Green – A cult film starring Charleton Heston showing an environmentally decadent scenario of the future, in which world food supplies are reduced to human remains.

spontaneous healing – Instantaneous healing of disease, unrelated to drugs and surgery, not understood by mainstream medicine.

Star Trek – Popular TV show depicting life in space in the future.

Star Wars systems – The U.S. Government's ballistic missile defense system.

string theory – An attempt to formulate a unified field theory within mainstream physics. This approach is mathematically complex, invokes dimensions beyond time and space, but doesn't address the phenomena of consciousness.

strawman – A hypothetical scenario which is almost certainly subject to revision.

suppression syndrome – A Twentieth Century phenomenon which is blocking the timely dissemination of new science and technology.

synchronicity – A word first coined by the late Swiss psychologist Carl Jung, these are meaningful coincidences.

systems theories – Theories which bring in a wide range of disciplines in order to address the whole picture.

Teams of Light – People who band together to bring in the new paradigm and the Consciousness Revolution.

telepathy – Communicating with the mind, beyond the five senses.

Tesla, Nikola – The great Yugoslav/American inventor who one century ago was the foremost pioneer of free energy.

ufology – The science of UFO phenomena, including sightings, ground traces, animal mutilations, abductions and other encounters with aliens.

Unified Field Theory – Physicists' attempts to reconcile the four known forces of nature into one equation. Such a theory has so far been elusive to traditional approaches, seems to require the existence of dimensions beyond time and space, and is mathematically complex (e.g., string theory).

unipolar generator – A rotating magnetic disk generator, sometimes called an N–machine or homopolar generator.

vibuti – Holy ash.

vortex – A whirling motion or mass; a destructive force that draws one in.

zero–point energy – Free energy, which comes from the "zero–point field." (ZPF is discussed in Chapter 11.) Called such because of the discovery that, even at absolute temperatures of zero degrees, the energy is still present but it is difficult to detect or tap because its properties are the same everywhere and in all directions.

Zeta Reticuli - A star in the Southern Hemisphere. Some UFO abductees claim the gray aliens come from Zeta Reticuli, although good evidence is sketchy.

Afterword

We found an attractive Maui beach home after a month of searching. The house miraculously surfaced once a much less adequate place we thought we had claimed was suddenly and mysteriously taken away from us -- another emotional roller coaster that came out positive. The move out of our former Oregon home had been even more laced with challenges in getting the packing done and lining up all the stages required of a shift from the remote winter woods to a faraway island which had felt to be only a dream. The logistics of moving a grand piano, furniture, framed art and 250 boxes up hill over 27 to 50 stairs, sometimes in heavy snow, rain and sleet, was as challenging as we had ever experienced.

But it was worth it ... Hawaii is the fulfillment of a lifelong dream expressed in retrospective scenarios such as "2020 Hindsight" which began and ended *The Second Coming of Science.* I had often thought I would move to the islands well before my death. This dream is now being fulfilled.

Throughout all the moving, traveling, book finishing, art shows, etc., it is a wonder Meredith and I are still talking to each other. We will now need to become reacquainted with our more refined selves, as we retreat to our new nurturing space for cocreating more miracles in the void. In spite of all the seemingly impossible obstacles of our time, I do see hope for us and for the Earth. The changes and adventure will be unprecedented.

I also took two recent trips which deeply bear on *Miracle in the Void.* The first was a meeting hastily convened by philanthropist Laurence Rockefeller on the subject of credible witnesses (astronauts, military and intelligence people) coming forward to testify on their UFO experiences. The 25–person conference was held in Asilomar on the Monterey Peninsula, a location which was not disclosed to me until just before the trip. To keep things confidential, they sent me an air ticket written to a pseudonym.

The intrigue was sharpened when I met and heard the private sharing from those who had the courage to come forward. Our goal is to obtain an executive order to grant immunity against prosecution for violating secrecy oaths in testifying. Mr. President, are you listening?

The people involved are all highly professional, and their stories are sincere, truly beliveable, and at the same time extraordinarily revealing and consistent with secret U.S. (and U.S.S.R.) Governmental involvement with alien cultures. Are we close to the moment of revelation -- the paradigm trigger itself? Perhaps so.

I also had the opportunity to become re-acquainted with Apollo 14 astronaut Ed Mitchell, who like me, did not come to testify as a witness (Ed hadn't seen anything alien on his journeys). We served as sort of ombudsmen, hearing the reports and asking questions of witnesses. Not only had the two of us been involved in the astronaut program and were founders of New Science organizations (Ed had founded the Institute of Noetic Sciences and I had co-founded the IANS). During our reunion, we also discovered that we were thinking alike about the theoretical understanding of free energy and consciousness, which I described in Chapter 11. And he too is finishing up a book on all this. Another synchronicity?

On the second trip , I flew over to Japan for a three-city speaking tour which was also breathless in its pace. Symposia sponsor Tetsuhiro Aso had also invited Canadian inventor John Hutchison to demonstrate his hew "crystal converter" free energy device. John has been able to produce a few tenths of a watt continuous electricity in up to a month of operation. These generators should be easy to build and scale up. John and I left quite an impression in Japan, which seems to be ready to break through the credibility gap into free energy. I expect to keep returning there.

I was recently struck by the research of John Wren-Lewis as reported in Marilyn Ferguson's *Brain/Mind Bulletin* (May, 1995) and the *Journal of Transpersonal Psychology*. The author describes his near-death experience (and those of others) as an unprecedented surrender to the moment, a letting go of our human survival defenses and any anxiety about the future, "a shining dark" formed in the back of the head. "It was a kind of blackness," he wrote, "yet the absolute opposite of blankness, for it was the most alive state I'd ever known -- intensely happy, yet also absolutely peaceful."

Perhaps our potential ability to let go so completely *is* our miracle in the void.

Brian O'Leary
Hana, Maui
July 12, 1995
full moon

Appendix I

Context and Guide for Using This Book

"The resistance to a new idea increases by the square of its importance."
–Bertrand Russell

As an Ivy League physics faculty member many years ago, I had thought myself to be at the pinnacle of humankind's knowledge. My belief system in mainstream science had been well–defined and restrictive. Then I might have dismissed with a sentence or two that there was no such thing as free energy or a dynamic ether that could provide power from the vacuum of space. I would have echoed the sentiments of my colleagues, all of us ensconced in our ivory towers.

How wrong I would have been. Only during the past three of my fifty–five years would I have admitted of the possibility that we have scientific evidence that anomalous energy production is not only feasible, but it promises a nonpolluting future. When this truth began to sweep across my mind, a Pandora's Box began to spring open at the deepest levels of my being.

My discoveries about all this led to a desire to write a book about the human quest to develop unlimited clean energy. This project was originally to be a rigorous study of underlying theoretical concepts and a status report on free energy experiments that have commercial potential. Cold fusion, rotating magnet generators, solid state devices, Tesla technologies, and other systems could soon replace power plants and oil–burning internal combustion engines as our mainstay energy.

I got only so far. A year ago my research took a curious turn at midstream. The positive results of both theory and experiment exceeded my most optimistic projections. I acquired more than enough proof, experience and relationships to last a lifetime. As I began to deepen my understanding that all this was very real, I became both excited about our potential but shocked at the extraordinary degree of suppression of the technology for nearly a century—whether by isolation, confiscation, threat, ridicule, or lack of funding and colleagueship. I found it difficult to hide my anger in looking at and writing about this subject.

I also began to ponder the implications of transforming the world energy economy of $2 trillion annually, how we have been avoiding the truth about our free energy potential, how we still insist on abusing the environment and mortgaging our futures with obsolete energy technologies. I visualized a weary and wary Earth trying to adapt to global warming, an issue raised again by scientists because of a strong and persistent El Nino condition in the South Pacific Ocean, and the potential spread of airborne disease from the resulting weather extremes. Among other symptoms of this historically unprecedented situation the plankton population, lying near the bottom of the food chain, is dying out in parts of the Pacific. Reversing the greenhouse effect can happen only when we radically reduce the amount of fossil fuels we burn from internal combustion engines and central station power plants.

And we have the foul air, oil wars, supertanker spills, etc. Nuclear power seemed to be no answer either, after Chernobyl; "hot" fusion keeps running into dead ends; and solar and solar-derived sources suffer from being diffuse, intermittent and economically challenging. Hydrogen fuel is one possible solution if we can find a way of producing it inexpensively—for example, through free energy. We seem to be like sheep under power lines unable to act in the face of unprecedented pollution worldwide.

I also saw the other side of the issue—that this technology is potentially dangerous and could be used for weapons and is therefore an issue of "national security", inviting (by analogy to nuclear energy) an unwritten public policy of secrecy and suppression. But can we afford to any longer hold Earth and all her species hostage simply because we are too immature to apply the very technology we need to get out of our mess? The very existence of free energy would seem to show we must grow up. I began to think, why hasn't this been debated publicly?

I soon discovered that we as a culture seem to be obsessed with a resistance so glaring, even Bertrand Russell may have been surprised, perhaps until he gave his formula, cited in the opening quote, a more careful look. For the stakes on this one are very high.

I began to see that the denial we harbor about free energy is similar to that directed toward other new science topics I investigated in

Exploring Inner and Outer Space and *The Second Coming of Science*—including research on UFO sightings, ground traces, animal mutilations, human abductions, Mars anomalies, crop circles, alternative health and healing, psychokinesis, precognition, biocommunications, survival of death, reincarnation, etc.

Each time I came upon an unknown truth that could shake us to our very cores, I found myself going through a curious cycle of feelings and behaviors that followed my initial realization that lifted me out of denial. I later discovered this progression to be a grieving process for the loss of familiar beliefs. I've been grieving my departure from that world view for some time now. I found that other scientific and medical expatriates are grieving too. I began to see similarities between these feelings and those identified by Elisabeth Kubler-Ross in her seminal research on grieving an impending death and the loss of loved ones. As increasing numbers of us become more aware of some of the coming radical changes, could it be, as the shock of change becomes evident, we will all need to be moving beyond our denials and into a grieving process? Is this our fate, as the ultimate revolution inevitably unfolds?

My work suggests a strong yes to that question, giving our futures a (subjectively) predictable psychological pattern. The structure of chapters in this book, as well as disciples represented in Meredith's companion paintings *The Last Supper of Gaia* and *The Resurrection of Gaia*, follow the stages of grieving of the old paradigm as well as celebrating the stages of transformation into the new. I began to experience and relate how it feels to grieve and transform, perhaps as an advance scout into an unknown that becomes a known.

Then I looked at the political aspects of the coming energy revolution, especially regarding the three E's—the economy, energy and the environment. I poignantly remembered the time when, as a consultant to U.S. Congress, I studied the origins and potential solutions to the energy crisis of the time.

My research at that time focused around the enormous ecological problems we had wrought on ourselves by abusing our energy sources—particularly oil. But when the OPEC cartel's grip loosened two decades ago and the energy crisis faded into lost memory, we have since *tripled* worldwide oil consumption rates and will have depleted this precious resource from the crust and face of the Earth in less than forty years!

I also had to take a fresh look at my chosen profession in the physical sciences. How complete are we, really, in knowing our universe? Why haven't we gone deeper than quantum theory in considering the unexplained effects of our consciousness on the material world as well as the existence of free energy itself—phenomena which have been ignored by most of our scientists? Why haven't the implicate order of David Bohm; the dynamic ether of Harold Puthoff, Edgar Mitchell, Harold Aspden, Bruce DePalma, Paramahamsa Tewari and Shiuji Inomata; the scalar energy models of Tom Bearden; the order-from-chaos theories of Ilya Prigogine and Moray King; and the psychokinesis experiments of Princeton's Robert Jahn—all eminent scientists—been incorporated into discussions among physicists?

Why hasn't biology taken seriously Rupert Sheldrake's ideas on morphic resonance or Cleve Backster's experiments on biocommunications (telepathy)? Why does mainstream medicine continue to rely exclusively on drugging and cutting when the mind-body medicine of Deepak Chopra and others are making great strides in curing disease?

And why haven't our philosophers or religious leaders truly looked at the scientific revolution anew, incorporating updated facts into formulating expanded world views that accurately reflect our increasing awareness of the power of consciousness? Why haven't we looked at the broad sweep of history to see that our current materialism, reductionism, determinism and anthropocentrism are on a collision course with a deeper metaphysics which has been held at bay by a money-driven culture?

I found that simply by revising the direction of focus within these four P's—physics, psychology, politics and philosophy—we could quickly come into an age of a sacred science, whose potential is staggering to contemplate and experience. This book project catapulted me into unexpected territory, far vaster than any technical or reportorial piece on Tesla technologies or UFO sightings. My research, travels, talks and writings inevitably gave me a wake-up call to look at the big picture, much as the Club of Rome had in its oft-quoted dire predictions in *The Limits to Growth*, published nearly a generation ago.

I had to dig deeper into the essence of why and how we have consciously and unconsciously abdicated to an elite few the control of our resources, priorities, and destinies; how we have trusted others to keep secrets from us; how we have allowed them to suppress promising new developments that might threaten their own vested interests; how we have wasted resources and polluted needlessly; how we are on the verge of global ecocide with little to show for preventing it from our governments or industries.

I confronted the distinct possibility that the Western world will economically slide ever further as Japan begins to manufacture free energy devices which could quickly supplant traditional sources. The

Department of Energy, the utilities, the oil companies, the nuclear industry and all the related infrastructure would become obsolete. What would all this mean?

As my knowledge further expanded I began to develop some unusual perspectives in embracing startling new scientific results. I also began to look at the wonders of the new paradigm knocking on our door, getting fleeting glimpses of our transformation from creativity to enlightenment to empowerment to joy and transcendence. Identifying these phases of transformation into the new, while not as strictly based on research as the phases of grieving, reflect the progression of my own experience as well as that of others.

The emotional, mind and heart-expanding aspects are not necessarily what you might expect coming from a traditional scientist, correspondent or historian. Fortunately, many of these jobs are being superbly done by others. The explosively growing scientific and engineering results of free energy research, aided by the tools of the Information Age, are being well-documented in recent literature in the monthly *New Energy News* and in the proceedings of the annual symposia of The Institute for New Energy, a new organization in which I have played a part as a founder and board member (see resources section).

Veteran journalist Jeane Manning reveals the stories of suppression and promise of free energy in the cases of dozens of inventors whom she has interviewed over the last thirteen years. I believe her new book *The Coming Energy Revolution* will be a classic in the field, adding a contemporary context to the many writings by and about Nikola Tesla, T. Henry Moray, Walter Russell, John Keely, Viktor Schauberger and other pioneers of this technology.

The Paradigm Shift

About a year ago, my project shifted away from the technical/historical aspects to a look at the broader implications of introducing this technology. I soon concluded that a "megaparadigm" (not megatrend) is knocking on our door, that the changes reach into our very cores, so deeply that not only will our thinking change, but our behavior, our religion, our philosophy, our politics, law, education, and virtually all other aspects of our being.

A logical approach to testing the hypothesis that we will be experiencing radical changes soon is to examine deeply what is happening on our scientific frontiers such as free energy, alien contact, healing with the mind, psychokinesis, etc. Any one of these is a candidate for being a major paradigm trigger. History provides ample evidence that revolutionary new developments in science and technology are both triggers for paradigm shifts as well as candidates for resistance, as Bertrand Russell so wisely stated.

Take the case of free energy. About a year ago, I became convinced as a scientist/futurist/author that we have proof-of-concept shown by numerous researchers plus some sound theoretical work. I noted that the probability of potential commercial success and worldwide dissemination (say, within ten years) was nearly 100%. Too much was going on in too many places for the seeds I have seen planted not to sprout and flourish as commercial opportunities.

Therefore, our worldwide energy infrastructure would tumble. Money being the main power of our time, we can take Russell's statement to mean that the suppression of free energy would now be staggering, whereas the resistance to the latest trend in existing industries (computers, for example) would be relatively small.

Let us take a glimpse of how enormous the coming shift will be. Consider the potential resistance to free energy, where vested interests of $2 trillion per year are being threatened. Compare the scale of this to some larger individual projects such as new power plants, dams or U.S. federal R&D programs on hot fusion, particle accelerators, Star Wars systems, and various other weapons, aircraft and space projects. Some of my earlier work suggests that these undertakings typically reach a "pork barrel" threshold on the order of one-half to a few billion dollars per year, at which time they reach a period of stability. Each of these projects begins to have a life of its own, but only after some debate about whether it is needed. Some of them don't make it at all because of the resistance. The ones that do make it have nothing to do with the new paradigm.

A little mathematics shows that the resistance to free energy would be very roughly a *million* times greater than some of the bigger projects of our time. (For those of you mathematically inclined, according to the Russell formula, a factor of one thousand greater spending in vested energy interests of $2 trillion per year versus the resistance to typical, contemporary megaprojects whose annual budgets are $2 billion, would be squared to become one million times the resistance.) Looked at in this light, it is no wonder that we have so thoroughly suppressed free energy technologies during this century.

Similar reasoning might apply to the UFO phenomenon and alien contact. So radical and far-reaching are the implications, it becomes less of a surprise that the forces of suppression such as governmental denial and secrecy continue to carry the day, even in the presence of overwhelming evidence to the contrary. In the end the public may decide on the UFO question. The Cold War is over and Gallup Polls have consistently shown a majority of Americans support the alien interpretation of these phenomena.

Perhaps scientists authoring a 1961 Brookings Institution report knew about this when they stated flatly that social unrest, national insecurity, and the careers of technocrats, financiers and scientists could be seriously threatened, if the release of any information about alien contact and/or artifacts were to be made public. Therefore such information should not be released. My experience in researching blatant governmental coverups and academic avoidance of such "touchy" topics suggests that the Brookings report has been adopted as federal policy to help keep the ruling elite in the saddle. Yet the credibility gap about the UFO question remains...until, as we explore in this book, a development such as free energy triggers the rest of the new paradigm.

Thought Exercise

Over the coming minute or two I invite you to do the following thought exercise. Envision if any one or more of the following were to happen as a real experience for most all of us: (1) contacts with aliens, (2) widespread availability of free energy, (3) pervasive use of mind–over–matter for healing and other applications, and (4) knowing much more of our essence by transcending death, Earth, space and time. Pick the one(s) that you might believe to be most credible and might come soon.

Now imagine that we, the public, supported efforts toward achieving breakthroughs and widespread education and implementation in your area(s) of interest. What would happen?

If your reflection on these ideas stimulates you, I invite your further examination. My own direct observation of these phenomena as an independent scientist over the last sixteen years suggests a high probability that at least "one of the above" will soon be felt as a paradigm shift in science and technology. In time, these developments are sure to reach very deeply into our psyches. Think about it. Think about the ripple effect from one major development such as cheap and clean energy. Or what if it becomes widely recognized we are not alone in the universe, that we can heal ourselves, that we may transcend death?

If your answer is "no" or "unlikely to happen" to any of the questions, I invite you to examine the evidence presented here and elsewhere. But if it is "yes" or "likely to happen" for any of the others, the basic principles we will be examining remain the same. I will show the basic principles we will be examining remain the same. I will show how any *one* of the above will almost certainly trigger some or all of the others, with additional changes so unpredictable and massive, reality will never again seem to be the same.

I am optimistic something will happen soon. Another way of looking at these questions is whether, over the next decade, none of the above would happen. (If, for example, the probability that free energy were not to become available in ten years is one in four and the probability we will find out little or nothing more about alien contact during that time is one in two, the likelihood that neither will happen is one in eight. For either or both to happen, the probability becomes seven-eighths.) The probability for "one of the above" goes even higher when we examine other paradigm triggers, such as healing, psychokinesis and the survival of death. Then, like dominoes, the others might fall too in an atmosphere of new openness toward these "taboo" subjects.

Some of you may still be "turned off" by the context of my remarks, that they go too far outside an anticipated scope of free energy, and therefore compromises the credibility of the work because so many other controversial concepts are weaved into the discussion, such as UFOs, radical ecological and philosophical views, and personal subjective experiences. Dealing with so many subjects and jumping between them may sometimes feel disconcerting. I might lose some of an audience who expects more linearity or objective Cartesian consistency. To those of you of that persuasion, I suggest a skimming of this book, which reads more like a series of essays on the impact of personal and global transformation. In the process, I hope you can find a gem to work with.

A few years ago I myself would have been initially skeptical of this kind of presentation. As recently as three years ago, I knew little about free energy, and therefore was doubtful about its existing or impending reality. I had to study the subject intensely to know that it will be happening soon. Three years ago, I was relatively well-versed in UFO studies and thus would have given the pervasive alien contact scenario a much higher credibility rating than our own potential to develop free energy. Events are overtaking our agendas.

The reason for my unorthodox approach is to search, from as many perspectives as possible, for metaphors for the new paradigm, for the new sacred science which must be invented to come to grips with our emerging perceptions of reality in which a traditional look would be inadequate. I have often taken the liberty of going more with a flow of feelings and intuitions than a traditional delineation of the subjects themselves, because these are often clues to our perceptions of the truth. Moreover, research on paradigm shifts suggests that the subjects themselves will be dissolved and redefined in ways we cannot anticipate. Exploring the emotions can provide the needed intuitions to develop a bold new plan of providing a sustainable future for the Earth. Our heart-knowing, our personal myths become the working hypotheses for the new sacred science.

Of course, you can easily skip over the segments you might not be interested in; all I ask is an open mind to investigate some of what I, as a scientist, have discovered. On the other hand, many of you may already know about some of the technical, historical, psychological and personal foundations I present over the first few chapters, which really only warm up to the substance of the book. I suggest you skim over those chapters.

During the course of the book I ask you to look at your own pathway to the new paradigm. What are your thoughts, feelings, behaviors, and expectations around the shifts which surround us, once enough of us step out of our denials? How can you make your own paradigm shift as we make our collective shift? What might become your part in an emerging debate about our future priorities?

I argue that most principal paradigm triggers have common origins and are interrelated. As soon as one old idea falls many of the others will fall too, analogous to the Industrial Revolution which ushered in this century with light bulbs, telephones, automobiles, and airplanes. Common denominators for the millennial paradigm include a physics which must include the "zero point" or free energy field (quantum fluctuations) as a factor in the creation of matter and energy, independent of Big Bang determinism. UFOs seem to have propulsion systems which can be most easily understood in terms of free energy and its closely-related concept, anti-gravity. Free energy, UFO propulsion, psycho-kinesis and mind-body medicine all share the same basic principle of a "new, new physics" through which we can create a vast array of unheard of realities through the power of our consciousness. I believe we will soon be able to develop technologies that transcend many of the limitations of time, space, matter, and misappropriated energy.

We have opened a Pandora's Box to a consciousness revolution and a sacred science. The genies of free energy and consciousness are out of the bottle, but our denial of the overwhelming evidence is still keeping the truth away from most of us.

Free energy and other paradigm-triggers will have a profound effect on every aspect of our lives. In this book I look at the coming changes from as many points of view as possible, to do a holistic technological forecast. In lifting out of my own denials, my personal odyssey and stream–of–consciousness become inseparable from the tidal wave of change. Extraordinary synchronicities provided clues to my personal awareness of how radical our new lives will really be, how we seem to inhabit two contradictory realities during these queasy mid–shift times.

Some of you may believe I may have gone too far with this, that I have been too intense and angry in my tone during the first few chapters, which describe what is deeply out of balance on our planet. But think about it. How would you react to the coming changes after some scientific and personal exploration? I hope this book will help you answer these questions.

Brian O'Leary
Ashland, Oregon
February 1995

Appendix II

Presented to the International Symposium on New Energy,
Denver, May 12–15, 1994

GREEN POWER:

The Coming Free Energy Revolution and the Return to Eden

by

Brian O'Leary, Ph.D.

Abstract

This is an energy policy and opinion paper about the potential for free energy to reverse many of the great stresses on our environment. Free energy could soon relieve us from burning up the Earth's oil in our transport vehicles, and stop us from abusing fossil fuels and nuclear power for generating electricity (air pollution, global warming, unhealthy, unsightly grid systems, nuclear waste disposal and proliferation, etc..) The existing worldwide $2–trillion–per–year energy industry will sooner or later give way to renewable sources, the most promising of which appears to be free energy. I have visited several highly qualified free energy inventors/researchers in their laboratories throughout the world and have witnessed several proof–of–concept demonstrations of free energy. Work so far has been suppressed in many ways. I propose establishing an international consortium in which the technical community could transcend the forces of suppression and greed, with the near–term goal of manufacturing appropriately–scaled free energy devices available for everyone. Free energy could become our chief catalyst for the forthcoming "paradigm shift" toward a New Science.

Introduction: Confessions of a Mainstream Scientist of the Seventies

Free energy. About twenty years ago, when OPEC raised its oil prices and energy crisis erupted, I began to look at how we in our culture were abusing energy from several perspectives. Over those years my professional roles included being a Ph.D. physicist/astronomer, author, astronaut, professor, science policy analyst, industrial aerospace scientist, Congressional aide and advisor/speechwriter to presidential candidates George McGovern, Morris Udall, Walter Mondale, and Jesse Jackson on science and energy issues.

During the public revelations about our energy abuse, I was a liberal academic of the tradition of a Berkeley, University of Texas, Cornell, Caltech, San Francisco State, Hampshire, Amherst, University of Massachusetts, Princeton and Penn. (Believe it or not, I had served on the faculties of *all* of these Universities between 1968 and 1980 until I discovered slowly that the academic world in my field was much too limiting.)

I also worked during the early 1980s for Science Applications International Corporation (SAIC) in a vain attempt to study advanced space energy concepts for NASA which they only supported at rare intervals. The only lucrative projects came from the U.S. Strategic Defense Initiative (Star Wars) program, which I refused to work on, so I had to leave the company in 1986, when I became a free-lance "new scientist" speaker/author poised on the edge of insecurity yet free to search for the truth (O'Leary, 1989 and 1993a).

This rather unsettled history resulted in a lot of exposure to scientific and engineering reality. It also perpetuated a myth most mainstream scientists seem to harbor: that radical new ideas such as free energy and UFOs are to be avoided at all cost. Therefore, I was (naively) protected from inquiries about fringe science.

Although rarely confronted about it from the outside, I had been influenced by some of my more vocal Establishment colleagues that any prospect that we could extract electricity, mechanical power, heat or cold from what we call nothing—the vacuum—was basically a nutty idea of a grade B, uneducated science fiction writer. My years of physics training and teaching had firmly grounded me in energy–mass conservation laws that appeared to hold us in their grip as the planet/universe seemed to be heading helter-skelter toward ecological/entropic disaster. The direction was due in large part to our abuse of energy sources—something about which I began to become acutely aware and depressed during the 1970s.

"You can't get something from nothing," my colleagues had smugly admonished me and I in turn had admonished my students. Perpetual motion was a dead issue. Otherwise, we high priests of reality continually argued, why don't we already have a machine that produces more energy than goes into it? Because such a device didn't exist, we surmised, we see continuing proof that there is no free lunch, that our sacred conservation laws are upheld.

In a physics lecture I once gave at Princeton, I even risked my own life "proving" to the students the energy conservation laws: I climbed to the top of a tall ladder in the front of a cavernous Gothic lecture room, and with my back pinned against the wall, I let go of a 20-kilogram weight at the end of a huge pendulum, which swung across the room from my nose and then dramatically hurtled back toward me until it came to rest a mere inch from my face—saved again by physical orthodoxy! (The demonstration showed energy conservation, with small losses in friction and air resistance, accounting for the fact that the pendulum ended up about one inch short of its release point—my nose.)

Regarding our big global oil burnup and abuse of electricity, I soon adopted a liberal academic view that we shall have to continue developing incremental, capital-and-materials-intensive, land-encroaching, intermittent and diffuse energy sources such as the sun, wind, ocean, hydrogen fuel, biomass, and even huge space satellites, to slowly alleviate our energy crisis. Meanwhile, a middle-ground policy emerged that we use even more of the Earth's oil and build some more nuclear power plants with increased safety standards, while developing small steps toward energy conservation and global sustainability.

This moderate view has gradually taken hold throughout the world over the past twenty years, during which the consumption of electricity has more than *doubled* (Flavin and Lenssen, 1994) and the use of oil has more than *tripled* (Lowe, 1994). Meanwhile I have gradually grown to realize that all this needn't be so, and that we have an opportunity to spawn a major international initiative to significantly help return the Earth to its former pristine state by using free energy instead of abused energy. I am sure our short-sighted energy-abusing juggernaut will puzzle historians of future generations.

The Energy Crisis of the Seventies

Do you remember the gas lines of the mid–1970s, the grim statistics of increasing oil scarcities, air pollution, oil spills, oil wars, prophecies of nuclear meltdowns, radioactive waste contamination, nuclear weapons materials proliferation from foreign domestic nuclear power programs (like North Korea now) and other predictions of doom and gloom? As we head toward the turning of the millennium, a time which many feel could be an Armageddon, whatever happened to the energy crisis? Whatever happened to the Club of Rome report on the limits to growth? Many of us can recall that as the 1970s further unfolded, the OPEC cartel began to lose its grip, oil prices dropped, and Ronald Reagan was elected U.S. President.

As if by magic decree, the mass media and public consciousness decided there was no energy crisis after all. The oil glut resumed and any perception that there was a challenge to develop alternatives seemed to drop out of public awareness to this very day.

During 1975 I was special consultant on energy to former Congressman Morris Udall's subcommittee on Energy and the Environment of the U.S. House Interior Committee. I wrote speeches and orchestrated hearings for eight months full–time for Udall while he was running for President. I helped him develop an energy policy not unlike what President Jimmy Carter began to implement during the late 1970s and the Clinton Administration perpetuates.

The policy acknowledges the grim statistics of a highly polluting and expensive energy future in the coming decades if we do not at least slowly move away from fossil fuels and nuclear power as the mainstays of our electrical power generation. We are also bringing in strict emission controls for automobiles and heavy industry. Yet underlying these positive steps has been an enormous blind spot and resistance to innovative solutions to the ongoing worldwide energy crisis.

Even some of the most innovative technical and policy organizations in the energy/environment field such as the Rocky Mountain Institute and the Union of Concerned Scientists (Brower, 1992) seem to be resigned to a change to more traditional renewable sources (solar, wind, hydrogen, biomass, etc.) and increased automotive efficiency. Incrementalism and reform may revive public trust through continuity and gradual improvement, but in the long run, neither the limited supply of fossil fuels nor the declining quality of the environment can be swept under the rug for much longer. The problems of the seventies seem to have become the even greater challenges of the nineties.

The Department of Energy's R&D as a Thin Edge of a Wedge

The fleeting public perception two decades ago that we did have an energy challenge helped to spawn a Department of Energy (DOE) that supports research and development in alternative energy sources. But much of the DOE was simply bureaucratic old wine in new bottles, combining existing special interests in both fossil fuel and nuclear power and weapons. On the positive side, the energy crisis of the 1970s gave new life to the environmental movement, which seized the opportunity to view the Earth–as–a–whole–seen–from–space as a symbol of the needs of our Mother Gaia on whom we so depend.

But our cultural memory seems to be short. Until this day, it is apparent to me that little true progress has been made to stop our abuse of energy and the environment. Rather than moving Manhattan Project–or Apollo Project–style into the future, we instead appear to have reinvented the wheel of vested interests in fossil fuels and internal combustion engines and nuclear energy, and we continue to supply electricity from large central power stations through an unsightly grid system that may be a major health hazard through electromagnetic pollution from power lines.

As an example of resistance to new ideas, I recall that during the late 1970s and early 1980s, while I was on the physics faculty at Princeton and then senior scientist at SAIC, some of us tried to interest the DOE and NASA in a concept that would use large solar power satellites (SPS) in space to beam energy to the Earth in the form of low-density microwaves. Receiving antennas would collect the energy which would then be transformed to alternating current that would be piped into the existing grid system, gradually replacing fossil fuel and nuclear power plants. The concept had been tested and proven. The environmental impact of this idea appeared to us to be less than that of the available alternatives (other than free energy, of which I had been unaware).

In spite of all the positive points, the SPS concept died a slow death in bureaucratic circles after a token spending of a few million dollars in studies and tests. It died not because of any lack of merit to the idea. Instead, it didn't seem to have the support it needed because of its radical nature and because it was contrary to the vested interests of the Government and the power and fossil fuel and nuclear industries.

Through the SPS example, it became very clear to me that *any* radical new idea in the energy field is in for tough sledding in the face of vested interests within the U.S. Government and established industry. The SPS concept was such that it appealed to the some of the utilities (base–load power would still go through the grid system from central–station antenna farms that replace fossil fuel and nuclear plants), but it certainly did not appeal to the oil or nuclear industries. Of course, free energy would appear to help none of the above groups and so has been suppressed.

Most of the billions of dollars of Department of Energy research and development funds are still spent each year to develop and promote the expanded use of fossil fuel and nuclear energy. In my years as a science policy analyst, I learned that government R&D projects form the thin edge of a wedge of great political and economic clout; today's blueprints are tomorrow's multibillion dollar realities (O'Leary, 1975). This leverage effect seems to play out once a project's investment goes over a billion dollars. Then the project becomes a new special interest, with contractors in most Congressional districts, etc. This guideline appears to hold regardless of the intrinsic merit of the project.

This is the main reason why the fossil fuel and nuclear industries are dominant. They got there first. It is interesting to see what lies ahead on the energy horizon, given the current "realities" of R&D activity. The largest single advanced R&D project in the DOE is the expenditure of over a billion dollars on the (still unfeasible) "hot" fusion concept that would involve building large power plants and more pollution from excess heat, radiation and power lines. Another significant portion of DOE funds is spent for high energy physics and weapons research not directly related to energy production. Much smaller funds are allocated to developing solar and other alternative sources and energy conservation. Nothing—not one penny—of American public funds outside the black budgets (which we don't know about) are invested to look at the source of energy which I believe *will* change the face of the Earth—free energy.

The Magnitude of the Global Energy/Ecology Challenge

In principle, free energy production will almost totally alleviate air pollution, global warming from carbon dioxide emissions, waste heat, Saddam Hussein's ecocidal fires, black skies, oil spills, acid rain, nitrogen dioxide, sulfur dioxide, hydrocarbon and ozone emissions, unsightly oil production and refining facilities, supertankers, gas stations, power stations, transmission lines, etc.

Free energy could also end our oil and natural gas thirst which is not only polluting the environment but is draining precious resources from the Earth at alarming rates. This lifeblood painstakingly formed over tens to hundreds of millions of years of gradual decay of organic materials within our crust, has been greedily extracted as if there were no tomorrow. I mentioned that oil production and consumption has more than *tripled* since the onset of the energy crisis. Almost half of the world's available oil and more than half of the natural gas have already been skimmed off the top of our best deposits and burned up willy-nilly mostly within one short human generation!

At present rates of consumption, proven U.S. oil reserves will last just ten years, and world oil reserves will last forty years (Brower, 1992). Even if these reserves were to prove to be twice as abundant as the estimates, we will run out of oil by mid-twenty-first century, with inevitable sharp price rises.

While all these facts may be obvious to the informed reader, they have been ignored by our energy policymakers. We are indeed borrowing the Earth from our children rather than inheriting it from our parents. *An ecological consensus is emerging that we must stop this and build a sustainable future.*

The economic impact of converting to free energy is also enormous. Revenues from the use of electrical power worldwide are now $800 billion per year (Flavin and Lenssen, 1994), a doubling over the twenty years that have passed since the energy crisis was articulated and then withdrawn from the world scene. This staggering cost even exceeds by twofold the size of the automobile industry, and is comparable to the annual income from taxpayers of a debt-ridden United States government. Spending on electricity is also comparable to the combined annual expenses of creating internal combustion and jet engines and the fossil fuels they use for transportation, heating and cooling, and other industrial systems.

Depending on how you measure it, the worldwide energy infrastructure that depends primarily on burning oil, coal, natural gas and radioactive elements consumes about $2 *trillion* each year, a figure so high it is hard to imagine the enormity of its grip on all of us. In the time it takes for you to read this sentence, the world is burning up more than one million dollars of fossil and nuclear fuels for end use in electrical appliances, heating, cooling and transportation systems.

We all know the enormous environmental toll: some effects appear to be irreversible even if we make the necessary changes soon. For example, we have not yet found sites for the safe disposal of long-lived highly radioactive waste from nuclear power plants and weapons. More irreversible damage such as increasing global warming, acid rain, noxious emissions, nuclear materials proliferation for weapons and Chernobyl-type events are inevitable if we do not change our ways.

Yet this very state of affairs has created an entrenched interest that has become so powerful that we seem to be blind to *any* new concepts, especially those which are as radical as free energy and cold fusion. We seem to be more interested in the controversy about whether or not these developments are real, rather than interested in seizing on a golden opportunity.

In short, we seem to have sunk into a false sense of security in continuing with an abusive energy infrastructure that is destroying the Earth and ourselves. We have created for ourselves an "electric jail", being increasingly boxed in by a gridlock of unsightly, unhealthy power lines and gas stations and the endless droning of internal combustion engines and other energy-related facilities that litter the landscape and the skies and the oceans.

If you think I'm exaggerating, may I suggest that you take a minute to look around you. Notice the effects of twentieth century industrial development related to our misuse of energy. Take a look outside. Do you see any power lines or hear or smell or see noisy trucks and cars? Could you visualize an environment without these polluting aspects?

Like the frog in the pond whose temperature is gradually raised each day until the pond becomes so hot he can't get out, we have gradually acclimated ourselves to our electric jail, gridlocked within it and forgetful of how enriching the more sustainable environments of one century or even twenty years ago felt to us. I live in the remote woods of the Oregon Cascades (now on the Maui coast) and can attest to the benefits of a peaceful, clean environment. Yet even here logging pressures have decimated the unique primary forests from which we used to feel some healing and nurturing.

Most of us have not faced the fact that we have abused our energy resources so badly and in such a short flash of history. We obviously can and must eliminate this Earth-battering if we are to survive.

Can you imagine a world without all these ugly, smelly, unhealthy and noisy machines that clutter and sprawl? Such a world becomes possible with free energy. But we will need to be wise in the ways this revolutionary source of energy is used. Can we direct the free-energy technologies toward environmental cleanup and enhancing the quality of life, instead of toward making more landfill mountains of consumer junk? New social structures will be needed to transcend the forces of greed, destruction and self-aggrandizement. Policies will need to be developed to facilitate the large displacements of employment and the economy. In this paper I will propose ways of evolving these structures.

Green Power: The Promise of Free Energy

The early promise of free energy is that our current polluting infrastructure will become obsolete. Electrical power systems probably based on compact solid state devices will replace the fuse boxes and circuit breakers of individual homes and buildings free of the grid system. The systems will also be portable, eliminating the need for storage systems such as batteries. They will replace the internal combustion engines in our automobiles and other transportation and industrial systems. And they can be used in the field to dramatically increase agricultural efficiency (e.g., pumping for irrigation) and therefore help eliminate hunger in Third World countries.

It appears that this source of energy, if properly applied, can go a long way to meet the most stringent ecological criteria of sustainability and the restoration of clean air, land and water. And perhaps most importantly, free energy will open a new paradigm of science and technology that will change the face of the Earth; it could make the Copernican and Industrial Revolutions appear tame. Redirecting $2 trillion per year from obsolescence to sustainability and affordability is unprecedented in world history.

But we will need to develop policies in which these energy devices are scaled appropriately to their end use, and not overused or applied to weapons.

In the longer term, I believe we will discover in a practical way that free energy can also be used for anti-gravity propulsion systems similar to those of UFOs. We will also open the Pandora's box that we can create and destroy matter and energy both with the power of our minds and with some smart engineering (O'Leary, 1993a). We will discover we can heal ourselves and others and can create our own realities. These are the technologies I believe will inevitably unfold with free energy being the most practical, immediate application.

So why don't we get on with it? It seems we have some deep-seated fears which are blocking the way, which haven't yet reached the public consciousness.

From Suppression to Revelation

It is remarkable to think that just one century ago, the highly polluting energy industries of the world were virtually nonexistent. It was also nearly one century ago when Nikola Tesla, the inventor of alternating current, was denied funding by financier J.P. Morgan for his research on free energy. The names of Tesla, Russell, Keely, Moray, Schauberger, T.T. Brown and untold others have paved the way for several dozen inventors worldwide who have shown, to the satisfaction of those scientists and engineers who wish to take the time out to investigate these matters, demonstrations of the concept that vast amounts of clean energy can be extracted from the vacuum of space by manipulating local magnetic and electrical fields in carefully–designed motors, solid state devices and chemical systems (such as has been demonstrated in cold fusion experiments.)

That such a concept is physically possible is explained in part by the theoretical works of Bearden, King, Puthoff, Aspden and others, as reported in last year's and this year's New Energy Symposia. I will not repeat these ideas here, but will mention that there already exists a substantial body of theoretical and experimental information to support the overall credibility of free energy. I have been surprised to see a breadth and depth of knowledge, dedication, and professionalism among leading theoreticians, experimenters and inventors in this field. Through my extensive travels worldwide, I learned that one does not need to make any *ad hoc* or metaphysical assumptions about the reality of free energy. I was most impressed with the intelligence and substantial achievement of many of the inventors and replicators of free energy devices.

My Visits to Free Energy Inventors and Researchers

Within the past few months I have I personally visited with Yull Brown, Floyd "Sparky" Sweet, Dr. Shiuji Inomata, Dr. Paramahamsa Tewari and others who wish to remain anonymous, in their laboratories. I have also met and conferred with several other inventors and researchers both at last year's and this year's conferences.

Although I cannot say with 100% confidence that I unequivocally observed free energy production, the tests I did observe are strongly suggestive of the effects the inventors claim to be getting. In my thirty years as an active (and cautious) scientist, I find the evidence for "proof–of–concept" to be as strong as it could be without actually using my own test equipment. Former Institute for New Energy Director Toby Grotz has suggested that experienced electrical engineers could make the necessary confirmations (or contradictions) of inventors' claims within one or two weeks, using standardized test equipment. This would be an important next step in free energy development.

Meanwhile, my basis for confidence in declaring my reality checks as valid is based primarily on observing repeatable, nonlinear electric outputs in many demonstrations and in replicated experiments which I have witnessed. I could not explain these anomalous results in traditional ways. These direct observations combined with a rudimentary theoretical understanding of the physics of free energy give me reasonable confidence that the effects both measured and calculated are real.

Add to this the fact that I am building relationships with these individuals based on growing mutual respect and trust among colleagues. I would be surprised that *all* of these people, for the years of work they have put into these experiments, are either deliberately or naively fraudulent. On the contrary, these are the explorers of a new reality, often cut off from the mainstream, because the mainstream will more often than not debunk this reality, with a denial based on the most superficial *ad hoc* reasoning.

Rather than the public's stereotypical image of the eccentric, out-of-touch garage inventor who is probably wrong, many of our free energy inventors and researchers are PhDs who are working in mainstream settings such as Shiuji Inomata at the Electrotechnical Laboratories in Tsukuba (Japan's "Space City"). Also President of the Japan Psychotronics Institute, he has been a full time government employee for the past 35 years. In India, Paramahamsa Tewari has a prestigious government position as Chief Project Engineer of that nation's largest nuclear power plants under construction. Both governments have permitted these two men to build their free energy devices based on Bruce DePalma's N-machine concept, something unthinkable in a DOE Lab in the United States.

The case of Japan is particularly interesting. Dr. Inomata has recently lectured governmental and industrial leaders (over 600 professionals showed up to his last seminar!), and Toshiba Corporation has recently invested $2 million to develop superconducting magnets for his new N-machine (see Dr. Inomata's paper in these proceedings). Being almost totally dependent on foreign oil for its energy and transportation needs, Japan has little to lose and a lot to gain from commercializing free energy. This could be another opportunity missed for other nations to come on board, with even more potential Japanese market dominance of products by their more far-sighted approach. When will we ever learn?

I base my impressions on an acquisition of knowledge about free energy which did not happen overnight. It has taken me months of learning and relearning time which few qualified scientists have bothered to take. We find that most of the most vocal naysayers have not spent much of their time addressing the puzzle of free energy. They are limited by peer pressure and funding and a strong bias against probing the unknown outside of their own specialties; I know, I was there!

Perhaps the most common error created by the debunkers is the erroneous assumption that if these machines were real, they would have heard about it. The history of science is replete with examples of leading scientists ridiculing (sometimes emotionally) new ideas because of this assumption, and later shown to be shown wrong (Kuhn, 1972; Westrum, 1982; O'Leary, 1989 and 1993a). This is, of course, about as far from science and rationality as you can get, and suggests that the suppression syndrome starts from its very apparent source—the scientists themselves.

Government officials and the media turn to the scientists for their information, and so also ignore the obvious. The result is an unfunded program revealed in the peripheral journals and alternative media. An example was the result of a recent in-depth interviews others and I had with a Washington Post reporter on free energy. What appeared in the newspaper was a light-hearted, "safe" historical piece on Tesla.

It seems that, by default, these Establishment mouthpieces define what is meant by credibility, which may actually have little to do with the real truth. One phone call from a mainstream journalist to Carl Sagan, for example, could quash a story. Alas, our most revered news sources do not have the final word on the truth. They are either suppressed or ignore the broader truth. Such is often the slow progress of science.

The Suppression Syndrome

So we might ask the question, why haven't we adopted free energy if we've had it for so long? Why have we put ourselves into electrical and oil gridlock and why have we so abused our resources and environment as a result? Why have we even gone to war over oil, which is so strongly a part of our involvement in the Gulf War, Somalia and even Viet Nam? Why have we misdirected untold trillions of dollars and sacrificed human lives and our surroundings and why do we continue to do so in the face of evidence that we can stop doing these self-defeating things?

How could so many decision-makers have kept free energy so completely suppressed from us so there is still not a single commercial machine (except perhaps the Swiss M–L Converter)? After all, the technology for availability is probably not that far beyond our reach—probably much less challenging of a project technically and financially than the Manhattan Project scientists faced in developing the atomic bomb, or than the Apollo Program scientists and engineers faced in sending men to the Moon, or than the Ph.D.s face working at the Princeton Tokomak hot fusion project which is still far away from the elusive "breakeven" point. Inventors have apparently been demonstrating free energy for mere thousands of dollars—not the billions and trillions spent on perpetuating more traditional approaches. Why has it taken so long for money to flow in logical directions?

In other words, how could the suppression of free energy have been so complete, so airtight, for so long? If our governments and scientists are ignoring the obvious, why haven't market forces in our cherished free enterprise system gotten wind of this and already briskly moved ahead? It seems that everybody is waiting for the other shoe to drop.

In my recent research on these questions I am coming to a conclusion I had previously thought to be unlikely: that the suppression syndrome pervades every aspect of any revolutionary new development. Usually the more radical the concept, the stronger will be the forces of suppression. Heading the list are the controversies around free energy, UFOs, the face on Mars, alternative healing and medicine, and mind–over–matter experiments (O'Leary, 1994; Eisen, 1994).

Somehow, most of us seem to not want to acknowledge the reality of these revolutionary technologies and to embrace their potential. In denying bold new possibilities, we seem to have been suppressing ourselves even when the suppression isn't coming from elsewhere. The Suppression Syndrome is so pervasive, so insidious, that its stranglehold on most all of us appears to be complete.

I believe the twentieth century will be remembered as the Century of Suppression and of Industrial Pollution. These patterns are so deeply ingrained into our psyches we don't quite seem to recognize them. But I believe, as the suppression stories come to light and new energy sources begin to come on line, the truth will be known.

Meanwhile, market forces are still resisting free energy because of certain social, economic, institutional and political idiosynchratic features of our time. Factors include:

• The U.S. and Canadian governments can halt work on patents and equipment under the Secrecy Order (Gag Rule) that stops free energy research in its tracks for some inventors such as Adam Trombly and Joseph Kahn. The governments justify these confiscations by asserting the device could be used for a weapon. They often stretch the truth under the guise of national security interests. And some inventors have received funds declared to be illegal under questionable securities laws, later to have their machines confiscated and/or themselves ex-patriated.

• Bizarre assassinations, gunfire, personal threats from CIA-types and industrial thugs have often displaced, discouraged, stopped and created paranoia and fear within a disorganized free energy community (Eisen, 1994).

• Our culture is incredibly short-sighted in general, ranging from the scientists themselves to the DOE and to the Harvard Business School "lapis pig" approach that optimize profits for a few greedy individuals, who through ignoring the obvious, can hold back what's needed for the world.

• Most inventors and researchers are underfunded and are hesitant to consummate their patents because they are still far enough from commercialization to want to delay patent publication only when they are near production. An inventor working in the absence of protection offered, for example, by the lawyers of a large corporation, risks losing a prematurely publicized patent to a competitor who basically steals the patent and offers minor improvements. And a small inventor would have to think twice of suing because of the great expense, time and effort of doing so. Lacking the needed scientific and engineering R&D support, it could take years or decades for an inventor to move from proof-of-concept experiments to manufacturing commercial models.

• Some inventors have been "bought out" in exchange for keeping their trade secrets under wraps, which then closes them off from sharing knowledge with those in the interdisciplinary teams that I feel will become necessary to develop this new industry.

Harold Aspden (1993) is more blunt about all this. "There is no way forward," he said, "for anyone involved in real research on free energy from ferromagnetism, unless that person understands the physics of the subject. The hit-and-miss ventures of those who build permanent magnet 'free energy' machines and get them to work anomalously only guides others equipped with the right training to take the research forward. I say 'only' because this is a simple situation. Those with the knowledge (mainstream scientists) do not want to believe that 'free energy' is possible. Those without the knowledge cannot prove their case, because they cannot speak the scientific language that applies. However, once on the scent and believing in what is possible but not knowing why, those 'experts' on magnetism will move rapidly in advancing the technology in the real commercial world."

In other words, the prospects of large financial returns from overoptimistic investment in a particular system kept under wraps is probably a misguided approach. We seem to have a horse race here, motivated by thoughts of the chance that a particular system might be the winner which could yield millions or even billions of dollars to lucky investors, whereas the other systems fail because of bad timing or underfunding and other suppressions. In a win–lose system in which even the syndrome itself is suppressed, it is no wonder we are spinning our wheels. In the Western world, the entire complex of scientific denials and industrial secrecy are gridlocking us. It's a crazy system! Meanwhile, the governments of India and Japan, with fewer interests invested in the current Energy Establishment, are supporting the efforts of some of their own researchers.

In summary, most free energy inventors, scientists and engineers have been underfunded, so progress is slow. The prospect of becoming a millionaire *a la* Thomas Edison by being among the first to develop a viable commercial model encourages secrecy and suppression. This all–or–nothing approach is not healthy for any of us and is keeping us away from what all of us and our planet so sorely need. I will be proposing win–win funding strategies that would virtually eliminate perhaps the greatest cancer of our time—the suppression. Because of our fear of the unknown, we are all suppressing what we need the most.

Challenges and Prospects for Free Energy

I suggest that there appear to be three main challenges for free energy: (1) **suppression** of all kinds has been efficiently blocking any move toward availability, and (2) the potential of free energy to replace existing infrastructures will cause enormous **displacements** in jobs, income and power to a degree which is unprecedented in our economy; and (3) the **abuse** of free energy technology could lead toward its overuse or as a powerful weapon.

I feel we cannot let potential abuse be a reason to stop or suppress the technology. We already have nuclear and chemical/bio-logical overkill.

Free energy is too important for the planet and for ourselves, and is inevitable. But *we must develop standards for the appropriate use of free energy to meet the strictest guidelines for global sustainability.* We need to be socially responsible creators, learning lessons from our abuses of nuclear energy, for example.

I do feel that once free energy devices pass the usual tests of cleanliness, cheapness, convenience, etc., the technology will quickly flood the worldwide marketplace. The Japanese already have shown an interest, and the consensus among free energy researchers I have talked with is that the day of first pervasive use is not far away (e.g., Grotz, 1994).

In the next two sections I will be making some recommendations on confronting two of the two major stumbling blocks to moving rapidly into a free energy economy: suppression and displacement.

Beyond Suppression

I recommend that we as a group come up with a resolution to work together to overcome the continuing atmosphere of suppression that has taken place. With strength in numbers, we could organize ourselves as a team to begin to develop appropriate commercial free energy devices more rapidly than could be done by one or a few researchers acting alone and in secrecy.

Some of this might include our willingness to share the financial rewards even if only one of us comes up with The Breakthrough. Such an organization would be neither a national government nor a particular company. Both are too tainted with perpetuating or gaining power, and the profit motive in and of itself obviously does not represent our collective interests for a sustainable future. We need a strong social mandate to transcend all this. For example, if any one company, country or agency were to take the lead, how are we to deal with monopolies, regulation, self-interest, etc.?

Because no existing institution would be able to manage the coming free energy revolution, it might be better governed by a new international consortium for free energy, whose members and labs would be charged with developing the appropriate technologies.

I sometimes go through the thought exercise, What could we do if we had plenty of money to support our activities? How would it feel to have funds given, no strings attached, for what you need, to work with an interdisciplinary R&D team that would proceed with free energy development? As in Japanese companies, the whole team gets rewarded for the support of the quantum leaps of the few. Nobody would get rich quick, but at least nobody with adequate qualifications and a desire to work on the team would be turned away or allowed to get poor or be a victim of suppression or isolation or outdated patent and securities laws.

What if we band together and draft a proposal that could attract beneficent funding sources ("angels") that could get this ball rolling? Imagine a friendly funding source that attached no rigid conditions on your work besides a commitment to help develop the technology for peaceful purposes.

Actually, this was my experience with NASA in its early years of optimism and growth during the 1960s and 1970s. It is amazing to think that, in the Apollo program, the fulfillment of a lunar landing took place only *eight* years after President John F. Kennedy set the goal. For me, Apollo was a wonderful experience of team spirit and positive attitudes. It was no wonder that Apollo was achieved within budget and ahead of schedule.

I envision a program initially funded on the order of $100 million or more to be used over the next few years to establish and run the proposed consortium, whose members would meet regularly, exchange research results, develop standardized test and replication procedures, set goals, discuss strategies, standardize terminology, share insights, share equipment and archives, document findings in other countries, interface with environmental researchers and with those studying the effects of the use of various materials such as magnets, work with communications people who will film and write about New Science, interface with educators from elementary school to university levels, plan future activities, discuss policies, socialize, and help to found free energy laboratories throughout the world to support free energy R&D for peaceful purposes. It would seem very wise for the DOE or, perhaps better still, a newly-formed public fund, to establish one or more laboratories in the U.S., analogous to the NASA centers in their early days and to Japan's Space City, which is already beginning to support free energy research.

The governing committee of this consortium would consist primarily of those individuals familiar with free energy issues and with high mind. They would come from a variety of disciplines, cultures and perspectives. The consortium would be primarily motivated to help our civilization emerge from the ashes of pollution. For this monumental task, we will need to move briskly as in the Apollo program.

Supporting Positive Change during the Coming Energy Revolution

I join many other futurists in foreseeing an inevitable and radical paradigm shift as a result of publicly acknowledged new discoveries in free energy, alien contacts, anti-gravity transport, alternative health, our eternal nature, etc. (O'Leary, 1993b). We all know of the profound effects earlier inventions have had on our lives—electricity, telephones, automobiles, airplanes, television sets, transistors and computers to name a few.

The developments on a near horizon will have an even deeper effect that transcends dollar values; *indeed Future Shock is here, and most of us appear to be uneducated about facilitating the inevitable displacements created by quickly rendering a multitrillion dollar industry as obsolete. Perhaps the greatest suppression of all is our (mostly unconscious) fear of the unknowns that await us on the other side of change.* Therefore, we want to deny the change for as long as we can, until the new perspective is so obvious we can no longer ignore it.

I have argued that, in our science and technology, we are at that watershed time, the time of paradigm shift, when we decide as a culture to move from one set of truths, or realities, to another. The old Newtonian view is beginning to go the way of the Flat Earth Society, and yet the prevailing wisdom in our conscious minds is still a Newtonian one.

Along with change comes the emotions of grieving an old world view. Well-established research on the grieving process suggests that soon most of us will move from our current denials toward anger, bargaining, depression and finally acceptance of the new (O'Leary, 1993b).

I am quite certain that as these revelations begin to shake the orthodox world and its delicate economic structures, many of us will experience a great deal of anger about the suppression issue and the fear of transition to the new. A bumper sticker reads, "The truth will set you free but first it will piss you off." I feel I have moved on to a cusp between depression and acceptance.

I believe we will all need to go through these emotions, so this is a good time to be tolerant and supportive of those making the changes. One good pilot concept are some of the efforts of some politicians to develop conversion programs for displaced aerospace workers who lose their jobs. Under such a policy, no jobs would be lost; they would be rechanneled to other priorities. I had helped to develop conversion strategies for various Presidential candidates. Some of these plans are beginning to be implemented in small ways as the Cold War fades away. I believe that conversion policies will now need to come closer to the center of our political thinking in all major endeavors of our culture.

Although these concepts are often politically unpopular, I am sure we will be hearing about conversion soon. The displacements are inevitable, and would be far greater if we don't plan now for a sensible energy future. We will begin to see that the "new truths" along with their powerful technologies can begin to lead the way to change. We will probably need a new political leadership or a newly educated existing leadership to acknowledge the truth and get on with the changes ahead.

Another function of an international consortium on free energy would be to devise questionnaires directed toward the public as to how they feel about these proposed changes to our culture. Such polls can then provide the needed basis to proceed with the changes and to give them added credibility.

The challenge is not one of whether or not Green Power is real. It is. Rather it is of our collective will to break free of our ignorance, the electric jail, the ecocide, the gridlock, the Newtonian rigidity, the greed and the vested interests.

While many of these thoughts and feelings are not new, we do sit poised for a historic opportunity. I am sure that among the great talent gathered at this conference, there will be breakthroughs and eventual mass production.

The real challenge that lies ahead is this: to take that brilliance to the world with love and for the benefit for all humankind. We need to develop that intention now, rather than sell it out to the forces of destruction. The stakes are high, now.

The energy sector provides a potent first example of radically shifting our priorities—from perpetuating powerful special interests to creating clean, convenient energy and sustainable environments. If we play our cards right, we can return ourselves to Eden, by embracing the new truths and by working as an Earth–team to transcend the ravages of hunger, disease, pollution, poverty and ignorance.

References

Aspden, H., 1994. "Experiments on Free Energy," *Nexus*, Feb.–March issue, page 50; originally a letter sent to Don Kelly.

Brower, M., 1992. *Cool Energy*, MIT Press, Cambridge, MA, p. 5.

Eisen, J., editor, 1994. *Suppressed Inventions and Other Discoveries,* Auckland Institute of Technology Press, New Zealand.

Flavin, C. and Lenssen, N., 1994. "Reshaping the Power Industry," Chap. 4 of *State of the World*, edit. L.R. Brown, W.W. Norton and Company, New York and London.

Grotz, T., 1994. "Around the Free Energy World in Thirty Days," *New Science News*, volume III, number 2, pages 2–3.

Kuhn, T.S., 1972. *The Structure of Scientific Revolutions* (2nd edition), University of Chicago Press.

Lowe, M.D., 1994. "Reinventing Transport," chapter 5 of *State of the World*, edit. L.R. Brown, W.W. Norton, New York and London.

O'Leary, B., 1975. "R&D: The Thin Edge of the Wedge," *Bulletin of the Atomic Scientists,* October issue.

O'Leary, B., 1989. *Exploring Inner and Outer Space*, North Atlantic Books, Berkeley, CA.

O'Leary, B., 1993a. *The Second Coming of Science*, ibid.

O'Leary, B., 1993b. "Sacred Science: Expanding the Box Even Further," *Proceedings of the International Forum on New Science,* edit. M. Albertson, IANS, Ft. Collins, CO.

O'Leary, B., 1994. "Survival vs. Suppressed Science," *New Science News*, volume III, number 2, page 1.

Westrum, R., 1982. "Social Intelligence about Hidden Events," *Knowledge; Creation, Diffusion and Utilization*, vol. 3, pages 381–400.

INDEX

About Kamapua'a Press

We established Kamapua'a Press in June, 1995 upon our arrival in Maui to settle here. This coincides with the occasion of printing this book as its first project. We will reveal the true meaning of choosing the word Kamapua'a in our second project. Until then ...

Signed copies of *Miracle in the Void* can be sent by priority (air) mail from the author for $25 postpaid at Kamapua'a Press, 1993 S. Kihei Road, Suite 21200, Kihei, HI 96753. Signed copies of Dr. O'Leary's New Science trilogy, including *Miracle in the Void, Exploring Inner and Outer Space and The Second Coming of Science* are available for $55 postpaid, air mail. His earlier books may be obtained in the same way for $20 each. Add $10 USD for foreign air mail shipments of one book, $20 USD for two or three books.

Dr. O'Leary is avaible to give lectures, seminars and workshops based on his New Science trilogy. For more information, write him at the above address.

About the Illustrator

Meredith Miller has been a full-time visionary artist for twenty-six years. Educated at the Art Center for Design in Los Angeles, she has produced hundreds of original paintings that have created healing and transformation for many of her collectors. For more information about available art and commissions, write her at the above address. Please inquire if you have interest in obtaining a hand-embellished graphic or poster of *The Last Supper of Gaia*. Enough inquiries will result in its reproduction.

More Great Reviews of *Miracle in the Void*

"***Miracle in the Void*** is one of the finest overviews of the `free energy' field I have ever read. Few authors have explained so well the depth of the suppression problem. Well done!"

-**Moray B. King**, Author of *Tapping the Zero Point Energy* and Free Energy Scientist

``***Miracle in the Void*** and Brian O'Leary's writings have boldly brought out the present limitations of our scientific knowledge and clearly point out the way for future progress on energy front globally."

-**Paramahamsa Tewari, Ph.D.**, India, Free energy Inventor and Director, Kaiga Power Project

"Brian O'Leary is a spiritual man with great wisdom."

-**John Hutchison**, Canada, Free Energy Inventor

"Just wanted to say 'thanks' for the copy of your latest book. As always, it's *really* good. And as you know, you sorta 'let it all hang out' and put a great deal of your inner self out there for all to see. That takes courage, a great deal of it ... You make a difference."

-**Tom Bearden**, Free Energy Theorist and author of *The Excalibur Briefing*

"Read O'Leary's book ... ***Miracle in the Void*** is an extraordinarily noble effort."

-**Bruce DePalma, Ph.D.**, New Zealand, Inventor of the N-Machine and former MIT faculty member

"***The Second Coming of Science*** and ***Miracle in the Void*** are feasts for the imagination. They are courageous, beautifully written books."

-**Michael Grosso, Ph.D.**, Philosopher, Author of *The Millennium Myth*

"I have been enjoying your book; you have a great capacity to draw metaphysics to earth and make visionary science real and alive!"

-**Alan Cohen**, Author and Lecturer